BROAD BAND

BROAD

BAND

The Untold Story of the Women Who Made the Internet

Claire L. Evans

PORTFOLIO / PENGUIN

Portfolio / Penguin
An imprint of Penguin Random House LLC
penguinrandomhouse.com

First published in hardcover in the United States by Portfolio, an imprint of Penguin Random House LLC, in 2018
This paperback edition published in 2020

Most Portfolio books are available at a discount when purchased in quantity for sales promotions or corporate use. Special editions, which include personalized covers, excerpts, and corporate imprints, can be created when purchased in large quantities. For more information, please call (212) 572-2232 or e-mail specialmarkets@penguinrandomhouse.com. Your local bookstore can also assist with discounted bulk purchases using the Penguin Random House corporate Business-to-Business program. For assistance in locating a participating retailer, e-mail B2B@penguinrandomhouse.com.

Library of Congress Cataloging-in-Publication Data

Names: Evans, Claire Lisa, author.
Title: Broad Band: The Untold Story of the Women Who Made the Internet / Claire L. Evans.
Description: New York: Portfolio, 2018. | Includes index.
Identifiers: LCCN 2017054620 | ISBN 9780735211759 (hardback) | ISBN 9780735211766 (epub)
Subjects: LCSH: Women computer scientists—Biography. | Internet—History. | BISAC: BIOGRAPHY & AUTOBIOGRAPHY / Women. | BIOGRAPHY & AUTOBIOGRAPHY / Science & Technology.
Classification: LCC QA76.2.A2 E93 2018 | DDC 004.092/2 [B]—dc23 LC record available at https://lccn.loc.gov/2017054620

ISBN 9780593329443 (trade edition)

Printed in the United States of America
10 9 8 7 6 5 4 3 2 1

Book design by Daniel Lagin

Illustration Credits:

Page 30: Grace Murray Hopper Collection, Archives Center, National Museum of American History, Smithsonian Institution
Page 45: U.S. Army Photo, courtesy of the University of Pennsylvania Archives.
Page 114: Courtesy of SRI International
Page 138: Courtesy of Stacy Horn
Page 147: Jim Estrin/*The New York Times*/Redux
Page 160: Courtesy of the University of Southampton
Pages 183 and 185: Courtesy of Jaime Levy

For the users

Contents

BROAD BAND

Introduction

THE DELL

When I was younger, I had a Dell.

It was a beige box fastened to the Internet with a 28.8K modem that screeched with every connection. Its keys were as tall as sugar cubes and slightly concave. The installation occupied the elbow of an L-shaped desk in my bedroom's inner sanctum. Over the years, I laid stickers in geological strata across the white laminate of my desk. Peeled one at a time, they'd have revealed earlier versions of the girl sticking them, like a candy passing through its flavors as it melts in the mouth. A teenage girl's room is a cockpit, an altar, and a womb: it contains her most sacred things, and it holds her as she grows, until eventually it ejects her into the world.

The Dell underwent its own changes. It ran every Microsoft operating system from MS DOS to Windows 95. The DOS era was wonderful: games on floppy disk, terminal commands. Over time, my monitor's blunt plastic bezel thickened with coats of glitter nail polish and Post-It notes. *GET A LIFE,* I wrote across the Dell's frame, in Sharpie, in anger, in devotion.

When the Internet came into my life, it was as though my monitor became a glass gate. It opened to an infinite channel. When the modem stuttered, I'd shower it in compliments: *You are such a good modem, and*

I believe you can do anything. It was my own compulsive folk tradition. I believed, then, that information, like people, needed support on its journey across the world. In my early years online, I learned how to write HTML and built rudimentary sites honoring my favorite bands. I sent passionate e-mails to estranged summer camp friends. I found answers to the questions I was too shy to ask. I made pen pals I was afraid of meeting. I journaled in pocket communities now obsolete. In short, I became myself, enjoying the freedoms the computer afforded me, freedoms both *from*—isolation, shyness, ignorance—and *to*—learn, experiment, discover, and play.

I abandoned the Dell when I left for college with a Sony VAIO, one of those tragic interstitial laptop models that will likely populate future museums of technology, with a detachable base that served mostly to heat the tops of my thighs. Like most consumer electronics in the United States, the Dell was likely landfilled, or else dispatched by container ship to China, Malaysia, India, or Kenya, where it was disarticulated like a chicken carcass, cables snipped, guts stripped of valuable metals and ores. Today I think about how the glitter-encrusted monitor must have looked to the underpaid laborers, working in a toxic field of unprocessed e-waste, who ground my Dell into plastic dust. Even once they've grown obsolete, computers never fully disappear—they only become somebody else's problem. Being mass-produced, they form part of our cultural memory, avatars, like my Dell, of childhood landscapes, or, like the Macintosh I never had, of personal computing as a whole. Doubtless this is why we so often consider the history of technology as a row of progressively smarter machines: from Chinese abaci to room-sized cabinets tended by pliant workers, from refrigerators with cathode-ray screens to ever-smaller incarnations of silicon and plastic, dwindling finally to the familiar handheld pane of glass. Anywhere along the line, it's tempting to eulogize the box. To point to one and say, "The people who made this changed the world." This story is not about those people.

This is a book about women.

It's also a book about the use of computers, real and potential. This is not to say that men make and women use—far from it—only that the technological history we're usually told is one about men and machines, ignoring women and the signals they compose. Female mental labor was the original information technology, and women elevated the rudimentary operation of computing machines into an art called programming. They gave language to the box. They wrestled brute mainframes into public service, showing how the products of industry could serve the people, if the intent was there. When the Internet was still an unruly assortment of hosts, they built protocols to direct the flow of traffic and help it grow. Before the World Wide Web came into our lives, female academics and computer scientists created systems to turn vast storehouses of digital information into knowledge; we abandoned those in favor of brute simplicity. Women built empires in the dot-com era, and they were among the earliest to establish and grow virtual communities. The lessons they learned in the process would serve us well today, if we'd listen.

None of this quantifies cleanly, which makes these women's contributions to computing difficult to catalog and even harder to memorialize. Although this book owes a debt of gratitude to the fine historical research it cites, I also drew from first-person accounts given by the women in these pages and from the fragmentary documentation characteristic of technological history: screenshots, chat logs, abandonware, outdated manuals, and eroded Web pages. I've done my best to explore what software artifacts remain, learning Unix commands and the social conventions of old-world online culture with the diligence of a student abroad. May the servers whir long enough to support more virtual tourism, because these places will become only more precarious with time. An irony: even as computer memory multiplies, our ability to hold on to personal memories remains a matter of will, bounded by the skull and expanded only by our capacity to tell stories.

There are technical women in these pages, some of the brightest programmers and engineers in the history of the medium. There are academics and hackers. And there are culture workers, too, pixel pushers and game designers and the self-proclaimed "biggest bitch in Silicon Alley." Wide as their experiences are, they've all got one thing in common.

They all care deeply about the user. They are never so seduced by the box that they forget why it's there: to enrich human life. If you're looking for women in the history of technology, look first where it makes life better, easier, and more connected. Look for the places where form gives way to function. A computer is a machine that condenses the world into numbers to be processed and manipulated. Making this comprehensible to as many people as possible, regardless of technical skill, is not an essentially feminine pursuit. Nothing is. That being said, the women I talked to all seemed to understand it implicitly and to value it as fundamental, inalienable, and right.

To live with a box that connects the world to itself is expansive, life altering, and even a little magic. But the box itself is still only an object. If not taken to pieces and recycled, it'll poison Earth for millennia, a permanence justifiable only if we believe what happens before the landfill is worthwhile. Spiritual, even. Computers are built to be turned on, cables are meant to be patched in, and links are made to be clicked. Without the human touch, current may run, but the signal stops. We animate the thing. We give it meaning, and in that meaning lies its worth. History books celebrate the makers of machines, but it's the users—and those who design *for* the users—who really change the world.

Women turn up at the beginning of every important wave in technology. We're not ancillary; we're central, often hiding in plain sight. Some of the most wondrous contributions in these pages bloomed in the grubby medians of the information superhighway. Before a new field developed its authorities, and long before there was money to be made, women experimented with new technologies and pushed them beyond their design. Again and again, women did the jobs nobody thought were important, until they were. Even computer programming was initially passed off onto the girls hired to patch cables and nothing more—until the cables became patterns, and the patterns became language, and suddenly programming was something worth mastering.

A few notes before we go. I take as a given in this book that sex is to gender as body is to soul. "Woman" means something different for everyone. There's no end to the ways in which it can be inhabited, and any loosening of the categories liberates a great many individual lives. That being said, women often share experiences, and particularly in environments where we are in the minority, it's nice to look for commonalities that can bolster our solidarity. One more: the history of computers is an alphabet salad. We'll meet ENIAC and UNIVAC and ARPANET and PLATO and the WWW. It can be difficult to read these acronyms without feeling like the past is yelling at you. Please don't despair. It's half the fun.

Onward now. My Dell is gone, its memory wiped. What remains of it isn't etchings on a hard drive but markings on a person: the user pushing symbols around. My memories of the Dell are like memories I have of family and friends. They're memories of time spent together, of journeys traveled. Memories of revelation and transgression. That's the miraculous thing about technology: it's never wholly separate from us. Just as a hammer strengthens the hand, or a lens the sight, the computer amplifies a person, extending the touch of even a teenage girl into the world. I am the computer, and the computer is me.

I won't be the last to feel this way. And I certainly wasn't the first.

PART ONE

The Kilogirls

Chapter One

A COMPUTER WANTED

It's 1892 in New York City. In January, an immigration processing center called Ellis Island opened for business. In March, in Springfield, Massachusetts, a YMCA instructor desperate to keep a class of stir-crazy youngsters entertained indoors hosted the first public game of "basket ball." But the winter is over, and it's the first of May, just shy of summer, just shy of the twentieth century. It's before the screen, the mouse, the byte, the pixel, and one hundred years before my Dell, but there's a strange notice in the classified pages of the *New York Times*.

A COMPUTER WANTED, it says.

This ad is the first instance of the word "computer" in print. It wasn't placed by an indiscreet time-traveler, someone trapped in the Gilded Age and jonesing for the familiar glow of their MacBook. It was placed by the United States Naval Observatory in Washington, DC, which was by then several decades into a mathematical astronomy project: calculating, by hand, the positions of the sun, stars, moon, and planets across the night sky. The observatory's directors were not in the market, that spring, to buy a computer. They were looking to hire one.

For close to two hundred years, a computer was a job. As in someone who computes, or performs computations, for a living. Had one been browsing the *Times* that May Day in 1892 and decided to answer the classified ad, they'd soon be taking an algebra test. The Naval Observatory job was cushy, relatively: those who lived nearby worked in a cozy, informal office in Cambridge, far from the observatory itself, which was perched on a bluff above the Potomac. They clocked five-hour days, charting the skies from individual tables by a roaring fire, pausing often to discuss the scientific ideas of the day. The rest worked from home, from detailed mathematical plans they received in the mail. Computing, as one historian has noted, was the original cottage industry.

Every day, these computers—much as computers do today—would chip away at complicated, large-scale math problems. They wouldn't do it alone. Our new hire would be part of a team: everyone crunching their share of the numbers, with some correcting their colleagues' work for extra income. With pen and paper alone, the Naval Observatory team would chart the skies, just as other computing offices throughout the Western world would advance ballistics, maritime navigation, or pure mathematics. They wouldn't receive much individual credit, but whatever the problem was, they'd have been instrumental in solving it.

Computing offices were thinking factories. The nineteenth-century British mathematician Charles Babbage, whose desire to calculate by steam led to important early developments in mechanical computing, called what the human computing offices of his time did "mental labor." He considered it work one did with the brain, just as hammering a nail is work one does with the arm. Indeed, computing was the grunt labor of organized science; before they were made obsolete, human computers prepared ballistics trajectories for the United States Army, cracked Nazi codes at Bletchley Park, crunched astronomical data at Harvard, and assisted numerical studies of nuclear fission on the Manhattan Project. Despite the diversity of their work, human computers had one thing in common. They were women.

Mostly, anyway. The Naval Observatory hired only one female computer for its Nautical Almanac Office, although she was by far the most famous among them: Maria Mitchell, a Quaker from Nantucket Island, who had won a medal from the king of Denmark before she was thirty for discovering a new comet in the night sky. It came to be known as "Miss Mitchell's Comet." At the observatory, Mitchell calculated the ephemeris of Venus, being, as her supervisor told her, the only computer fair enough to tackle the fairest of the planets.

Her presence as a woman in a computing group was unusual for its time, but it would only become less so. Maria Mitchell discovered her comet only a year before the Seneca Falls Conference on the Rights of Women, which was largely organized by Quaker activists. Her church was the sole religious denomination allowing women to preach to its congregations, and Maria's father, an amateur astronomer, lobbied aggressively for her accomplishments to be recognized. Before the end of the twentieth century, however, computing would become largely the purview of women. It was female mental laborers, breaking intractable problems down into numerical steps much as machines tackle problems today, who ushered in the age of large-scale scientific research.

By the mid-twentieth century, computing was so much considered a woman's job that when computing machines came along, evolving alongside and largely independently from their human counterparts, mathematicians would guesstimate their horsepower by invoking "girl-years," and describe units of machine labor as equivalent to one "kilo-girl." This is the story of the kilogirls. It begins, as the most beautiful patterns do, with a loom.

THE SPIDER WORK

The loom is a simple technology, but in the warp and weft of thread lies the weaving of all technologically literate society. Textiles are central to the business of being human, and like software, they are encoded with meaning. As the British cultural theorist Sadie Plant observes, every cloth is a record of its weaving, an interconnected matrix of skills, time, materials, and personnel. "The visible pattern" of any cloth, she writes, "is integral to the process which produced it; the program and the pattern are continuous." This process, of course, historically concerns women. Around looms, at spinning wheels, in sewing circles, in ancient Egypt and China, and in southeastern Europe five centuries before Christianity, women have woven clothing, shelter, the signifiers of status, even currency.

Like many accepted patterns, this was disrupted by the Industrial Revolution, when a French weaver, Joseph-Marie Jacquard, proposed a new way to create cloth—not by hand, but by the numbers. Unlike a traditional loom, singularly animated by its weaver's ingenuity, Jacquard's invention produced remarkably complex textiles from patterns punched into sequences of paper cards, reproducible and consistent beyond a margin of human error. The resulting damask, brocade, and quilted matelassé became highly coveted all over Europe, but the impact of Jacquard's loom went far beyond industrial textile production: his punched cards, which separated pattern from process for the first time in history, would eventually find their way into the earliest computers. Patterns encoded on paper, which computer scientists later called "programs," could meaningfully entangle numbers as easily as thread.

The Jacquard loom put skilled laborers, male and female, out of work. Some took out their anger on the frames of the new machines, claiming as a folk hero the apocryphal Ned Ludd, a weaver said to have smashed a pair of stocking-frames at the end of the previous century. We use the term *Luddite* now in the pejorative, to describe anyone with an unreasonable aversion to technology, but the cause was not unpopular in its time. Even Lord Byron sympathized. In his maiden speech to the House of Lords in 1812, he defended the organized framebreakers by comparing the results of a Jacquard loom's mechanical weaving to "spider-work." Privately, he worried that, in his sympathy for the Luddites, he might be taken as "half a frame-breaker" himself. He was, of course, not—and he was dead wrong about the spider work, too.

Even as Byron made his case, Jacquard looms were producing a quality and volume of textiles unlike anything the world had ever seen. The mathematician Charles Babbage owned a portrait of Joseph-Marie Jacquard woven from thousands of silk threads using twenty-four thousand punched cards, a weaving so intricate that it was regularly mistaken for an engraving by his guests. And although the portrait was a fine possession, it was the loom itself, and its punch card programs, that really ignited Babbage's imagination. "It is a known fact," Babbage proclaimed, "that the Jacquard loom is capable of weaving any design which the imagination of man may conceive." As long as that imagination could be translated into a pattern, it could be infinitely reproduced, in any volume, in any material, at any level of detail, in any combination of colors, without degradation. Babbage understood the profundity of the punched-paper program because mathematical formulae work the same way: run them again and again, and they never change.

He was so taken with the Jacquard loom, in fact, that he spent the better part of his life designing computing machines fed by punch cards. To describe how these worked, he even adopted the language of the textile factory, writing of a "store" to hold the numbers and a "mill" where they could be processed, analogous to a modern computer's memory and central processing unit. Numbers would move through Babbage's machines, coming together as thread becomes whole cloth.

Babbage's machines—the Difference Engine, a hand-cranked mechanical calculator designed to tabulate polynomial functions, and the more complex Analytical Engine—were so far ahead of their time that they're generally considered historical anachronisms. His mechanical designs required a level of technical precision never before attempted, although the British government, for whom mathematical tables were a point of national interest, was willing to try. It funded construction of the Difference Engine in 1823, with an initial grant of seventeen hundred pounds; by the time it wrote off the project, nearly twenty years later, having spent ten times as much, there was still nothing to show for what the prime minister had by then determined to be a "very costly toy," and "worthless as far as science is concerned," save some partial models and four hundred square feet of confounding schematic drawings.

The machines made Babbage famous—and perhaps infamous—but very few people alive in his time were mentally equipped to understand what they were supposed to do, let alone how. One of those people was Lord Byron's daughter, Ada. In her short life, she would make one thing certain: that the spider work her father had so disdained would proliferate, unstoppable, into the following century and beyond.

RAYS FROM EVERY QUARTER OF THE UNIVERSE

Ada's alchemy was peculiar. She was the child of a passionate yearlong marriage between Byron and a bright, mathematically inclined aristocrat named Anne Isabella Milbanke, or Annabella. Byron was, in a former lover's estimation, "mad, bad, and dangerous to know," his passions Romantic in every sense; Annabella, on the other hand, was so sensible and well-bred that Byron teasingly called her the "Princess of Parallelograms." The couple separated amid rumors that the louche Byron had a more-than-fraternal relationship with his half sister Augusta.

Amid the scandal of that separation, the last thing Annabella wanted was for Ada to inherit any of her father's wildness or to suffer as a consequence of his notoriety. To keep her daughter on the straight and narrow, Annabella began a rigorous course of mathematical instruction from the time Ada was four years old. Math—the opposite of poetry. Or so she thought.

Byron absconded to Italy shortly after Ada's birth. He never made her acquaintance, although he inquired after her often. "Is the Girl imaginative?" he wrote to Augusta, knowing full well that Annabella, who kept their daughter purposefully secluded, would divulge nothing directly. Byron died unromantically of the flu in Greece in 1824, when Ada was only nine. As he died, he called to his valet, "Oh, my poor dear child! My dear Ada! My God, could I have seen her! Give her my blessing!"

His body was returned to England by ship, and huge crowds gathered in the streets of London to see his funeral procession of forty-seven carriages. When Ada finally learned her father's name, she wept for him, although it doesn't appear that she or her mother held his legacy in high esteem—Byron's portrait, in their home, was concealed under heavy drapery until Ada was twenty. But his mercurial spirit was alive in her. "I do not believe that my father was (or ever could have been) such a *Poet*

Ada King, Countess of Lovelace

as I *shall* be an Analyst; (& Metaphysician)," she wrote to Charles Babbage later in life, "for with me the two go together indissolubly."

Ada's sharp analytical mind was inflected by a wild imagination. Prevented from a formal university education by her gender, she thrived under private tutelage. A precocious and very lonely child, she designed flying machines and marched around the billiard table playing violin. She was also frequently ill, prone to episodes of what was then called hysteria, and barely survived a serious three-year bout of measles, during which Annabella took advantage of her daughter's bedridden condition to double down on schoolwork. But Ada was indomitable, agitated, and charismatic, and when she outpaced—and in one instance, seduced—her tutors, she educated herself with books and through correspondence with some of nineteenth-century England's most illustrious minds.

She was only a teenager when she struck up a close friendship with the well-known scientist Mary Somerville, who would answer her questions and encourage her studies. The logician Augustus De Morgan sent her problems by post, only to be astounded by the power of thinking represented in her responses. Had she been a man, he marveled, her "aptitude for grasping the strong points and the real difficulties of first principles" would have made her "an original mathematical investigator, perhaps of first rate eminence." She did not shrink away from difficulty, and she had a peculiar way of learning: she questioned the basic principles of mathematics to drill down to their fundamental meaning and understand them completely.

Ada first met Charles Babbage when she and her mother went to see his Difference Engine, the first of his very expensive, very unfinished mathematical machines, in London. She was seventeen; Babbage was forty-two. He displayed the machine—a piece of it, anyway—in a salon where he hosted Saturday-night soirées that attracted the most prominent names in society: Charles Darwin, Michael Faraday, Charles Dickens, the Duke of Wellington. It wasn't long after Ada's ritual debut in court, where she had worn satin and tulle and made whispered pronouncements to her mother about the various dukes to whom she was presented: Wellington, she liked, and the Duke of Orleans, too, but the Duke of Talleyrand? He was an "old monkey."

Ada diligently made the rounds, but she held her social obligations in low esteem. She was, however, immediately mesmerized by Babbage's machine, a hulking block of interlinked brass gears and cogs. "While other visitors gazed at the working of this beautiful instrument with the sort of expression, and I dare say the sort of feeling, that some savages are said to have shown on first seeing a looking-glass or hearing a gun," wrote an onlooker, "Miss Byron, young as she was, understood its working, and saw the great beauty of the invention."

Not long afterward, Ada became Ada Augusta King, after her marriage to a sensible aristocrat a decade her senior, and then, three years later, her husband's peerage elevated, the Countess of Lovelace. By the age of twenty-four, she'd borne three children—one, a son, named after her father—and was managing her family's homes in Surrey and London, but she continued to study mathematics every day, and she remained fascinated by the Difference Engine.

She pleaded with Babbage to let her be of service to his machines. "I hope you are bearing me in mind," she wrote to him in 1840, "I mean my mathematical interests. You know this is the greatest favour any one can do me." Being a countess came with social obligations Ada found immensely distracting from her true passions; she wanted a professional path, a vocation, to practice mathematics in some useful way that might cement her legacy as her father's poems had cemented his. Her letters—to Babbage, to her mother, to her many friends—reveal a woman consumed by the crippling fear that she might not have the opportunity to make her mark on mathematics. She was certain of her own unique talents: both her immense reasoning faculties, drilled into her by her mother's homeschooling, and her "intuitive perception of hidden things," the legacy of her absent father. "I can throw *rays* from every quarter of the universe into *one* vast focus," she wrote to her mother, who worried she might be mad.

Ada had affection for her husband—she called him "my *chosen* pet"—but she devoted her mental life to Babbage and his machines. She became his acolyte, and then his mouthpiece. His iconoclastic way of thinking appealed to her; she admired the imagination of his inventions. Having been raised in isolation, under the rigorous tutelage of a mother hell-bent on curbing any trace of Lord Byron's poetical fancies, Ada felt validated by Babbage. Like her, he understood that the manipulation of numbers—the highest levels of mathematical thought—had profound metaphysical implications. That math was a form of poetry in itself.

But by the time Ada was married, Babbage had all but given up on the Difference Engine. Impressive as it might have been to the British society passing through his Saturday soirées, it was only a very complicated adding machine, churning out rows and rows of numbers using the method of finite differences. The Difference Engine could have been used to tabulate error-free mathematical tables, to precisely "calculate by steam" the sorts of problems human computers had by then been working through with only occasional errors for more than a century, but Babbage was no longer interested in anything so practical. He had a bigger idea.

The Difference Engine's precisely milled cogs and wheels stored thousands of numbers, but Babbage longed that they store *variables* instead—abstract symbols standing in for numbers. Such a machine could do much more than arithmetic. It would be capable of solving *every* kind of problem. He began to make plans for a second, far more ambitious engine, one that would make the conceptual leap from mechanized arithmetic to full-fledged general-purpose computation. He called it the Analytical Engine.

If the Difference Engine was ingenious, the Analytical Engine was brilliant. Had it ever been fully built, the Analytical Engine would have been able to multiply two twenty-digit numbers in three minutes. The Harvard Mark I, an electromechanical computer built in the 1940s using some of Babbage's basic computing principles, was capable of the same task in about six seconds, albeit nearly one hundred years later; today, my laptop does it in under a millionth of a second. But the Analytical Engine was not an electronic machine: it was a cumbersome mechanical thing, its cranks, rods, and spinning gearwheels designed to be powered by steam. The word "engine" is right: to an untrained eye, the partial model of the Analytical Engine currently on display at the Science Museum in London looks like something pulled from the belly of an old steam locomotive. It has the formidable and hulking physical presence of a bank vault.

It was a tough sell. After all the money it had wasted on Babbage's Difference Engine, the British government certainly wasn't going to spring for a new model with even fewer immediate applications, and Babbage had nobody to lean on: in his obstinacy, he'd made his share of enemies in the British scientific community. In the hopes of stoking interest in his machine, Babbage accepted an invitation in the fall of 1840 to go to Turin and share his plans for the Analytical Engine with a group of Italian scientists and philosophers. He hoped that "the country

A small portion of the Analytical Engine's "mill"

of Archimedes and Galileo" might prove more enlightened than his homeland, but things didn't go as planned.

Seated in Babbage's Turin audience was a certain Luigi Federico Menabrea, a young military engineer who would later become a diplomat, and then the Italian prime minister. Soon after the presentation, Menabrea wrote a detailed paper, *"Notions sur la machine analytique,"* for a Swiss journal. When the intellectually curious Ada came across the paper, she immediately began to translate it, correcting Menabrea's mistakes as she went. She presented the unsolicited translation to Babbage; impressed, he asked her why she hadn't just written an original paper, seeing as she was so familiar with the machine and its architect. The thought had not occurred to her. Babbage suggested that she should, at least, add some of her own notes to the translation. This compromise between modesty and intellectual ambition was amenable, and she undertook the project straightaway. But by the time they made it to the printer's office, Ada's notes—which she signed only with her initials, AAL—had taken on a life of their own. They were nearly three times longer than Menabrea's original text, and an order of magnitude more sophisticated.

In her notes, Ada synthesized the vast scope of Babbage's vision. It was no easy task: by the time he died, he'd dedicated thirty volumes of plans to the Analytical Engine. Enlivening her technical analysis with flights of metaphysical fancy, she aimed to make the machine comprehensible—and exciting—for an educated Victorian audience, particularly those among the scientific community and the British government, whom Ada and Babbage both hoped would come to their senses regarding the machine. Babbage was infamously stubborn and not a particularly good political player, and Ada knew his brilliance could easily be overlooked by those who found his temperament intolerable. "My dear and much admired Interpretress," he admitted.

But Ada didn't only explain the technical workings of the Analytical Engine. She imagined the impact it could have on the world, teasing out the implications of general-purpose computing to anticipate the transformative power of software. She understood that if the Analytical Engine manipulated symbols, then anything that could be represented symbolically—numbers, logic, even music—could pass through the machine and do wondrous things. "The Analytical Engine *weaves algebraical patterns,*" she wrote, using a textile metaphor, "just as the Jacquard loom weaves flowers and leaves." The possibilities were limitless, and hers was just the mind to articulate them: mathematically brilliant and poetically incisive in equal measure.

The work was taxing on her, mentally as well as physically. Like many patients at the time, she was prescribed laudanum for her maladies. Through an opiate haze, she labored in bursts of feverish energy between social appointments and periods of illness. Her mother disapproved of the work, and she tried to contrive family dramas to distract her, but Ada was tenacious. Correspondence between Ada and Babbage during this time was brisk and highly intimate. They sent letters back and forth across London, often several times a day. She chided him for his sloppy work, bristled when he edited her writing, and caught his errors, all the while referring to herself as his "Fairy," an apt description for the mathematical sprite she was. "That brain of mine is something more than merely mortal," she boasted as she sorted out all the ways the machine could deduce Bernoulli numbers. "Before ten years are over, the Devil's in it if I have not sucked out some of the life-blood from the mysteries of this universe, in a way that no purely mortal lips or brains could do."

The Analytical Engine would never be completed, but it represents the conceptual dawn of the computer age. The four components of its design—input, storage, processing, and output—remain core components of all computers today, and the strikingly original notes that Ada prepared to explain this new kind of machine would presage the literature of computer science by nearly a century. To demonstrate how the engine could calculate Bernoulli Numbers without any assistance from a "human hand or head," she wrote mathematical proofs that many scholars characterize as the first computer programs ever written, and all for a machine that never even existed. Although Ada had three children, she referred to her notes on Menabrea's essay as her firstborn. "He is an uncommonly fine baby," she wrote to Babbage, upon completing her draft, and "he will grow to be a man of the first magnitude & power."

It's telling of Ada's time that she characterized her work as male and signed her notes with only her initials. Although she was encouraged in her lifetime by high-profile supporters—Babbage chief among a circle that included her tutors, husband, and scientific friends—her path was decidedly unorthodox. Even her mother barely tolerated it. "Not even countesses," writes Sadie Plant, "were supposed to count." Beyond her friend Mary Somerville, she had few female peers, and her accomplishments required a dogged and persistent self-education, a near-manic dedication to mathematics that defied convention and damaged her health.

Ada had been prone to illness her entire life, suffering from bouts of dizziness, pain, fainting, and nervous malcontent. Her symptoms were written off as hysteria and managed with her regular doses of laudanum, which she anticipated eagerly, her eyes burning. At thirty-six, the same age as her father, Ada died of what really ailed her: uterine cancer.

She had all but given up on mathematics. In her final years, she bet compulsively on horse races, using her mathematical acuity to calculate odds for an ad hoc syndicate of male friends. One biographer has suggested that she hoped to win the fortune required to build Babbage's Analytical Engine, but she lost so often and so spectacularly that she was forced to borrow money from friends and pawn family jewels. By the time she succumbed to protracted bed rest in London, she had become more like her father—mad, bad, and dangerous—than any Princess of Parallelograms. Floating in and out of reality with doses of laudanum, wine, and chloroform, she echoed the family chord of recklessness and tragedy. "I do dread that horrible struggle, which I fear is in the Byron blood," she wrote to her mother. "I don't think we die easy."

Like her father's, Ada's work outlived her, although it would be nearly a century before it was properly recognized. It took until the beginning of the computer age, when the magnitude of their prescience became undeniable, for her *Notes* to be republished, in a British computing symposium; its editor marveled, in 1953, that "her ideas are so modern that they have become of great topical interest once again." Ada was lucky to have been born wealthy, noble, and relatively idle. Even without a professional path, she was able to educate herself, and she had time to privately follow her passions. Still, she could have done so much more, and it's evident that she wanted to. Many brilliant women—born in the wrong centuries, the wrong places, or hoping to make an impact on the wrong field—have suffered similar fates, and far worse.

Reading Ada's correspondence, I see someone I wish I could reach out to, across the centuries, and say: you're right. Nobody can see it but you. But you will have inheritors. Granddaughters and great-granddaughters. They will sprout up everywhere, all over the world, and work with the same dogged, unrelenting focus. Other people will keep getting the credit, until one day they won't anymore. And *then* your history will be written, a hundred times, by teenage girls at their desks in the heart of their kingdoms, on machines beyond your wildest imagination.

KILOGIRLS

By her insistence, Ada Lovelace was buried next to her father in a small church near his ancestral estate of Newstead Abbey. Her coffin, finished in soft violet velvet, bore an inscription of the Lovelace family motto, an axiom she'd embraced as her own while toiling over her notes on Babbage's Analytical Engine. LABOR IPSE VOLUPTAS, it read. "Labor is its own reward."

Labor would remain its own reward for a long time. By the end of Ada's century, although technically gifted women like her could find employment as computers on either side of the Atlantic, their formal titles weren't accompanied by commensurate status or compensation. In the 1880s, for example, the astronomer Edward Charles Pickering hired only women to analyze and classify stellar data for his Harvard lab, including his own maid, Williamina Fleming. Although he would later champion the women working in the observatory, even presenting papers on Fleming's behalf at astronomical conferences, Pickering didn't hire them out of advocacy. He'd just wanted twice as many workers on the job, given that women were paid half the going rate. "The Harvard Computers are mostly women," complained the director of a competing observatory, which employed only men, to a colleague, and they can be "got to work for next to nothing."

Known to history as "Pickering's Harem," the Harvard Computers cataloged ten thousand stars; Williamina Fleming, the erstwhile maid, discovered the Horsehead Nebula and helped develop a common designation system for stars, while her colleague Annie Jump Cannon could classify spectra at a rate of three stars a minute, and with a remarkable consistency that allowed her to discover a number of new and unusual stars. These women quite literally mapped the cosmos, but their wages were equivalent to those of unskilled workers—paid between twenty-five and fifty cents an hour, they earned barely more than they would have if they'd worked in a factory.

In the United States, the number of female office workers increased near the end of the nineteenth century, with a significant uptick after the American Civil War. Major wars have an unmistakable effect on gender and work, opening new employment to women; in this case, many were battlefield widows, looking to support themselves by helping to coordinate the affairs of an increasingly complex world. After the Civil War ended in 1865, as historian David Alan Grier writes, female computers were no longer "the talented daughters of loving fathers" as Maria Mitchell had been, "or the intelligent friends of sympathetic men," like Ada. They were "workers, desk laborers, who were earning their way in this world with their skill at numbers."

The First and Second World Wars, too, ushered thousands of women into the workplace as typists, clerks, and telephone operators, to say nothing of riveters. But it was the telephone companies that were the first mass employers of a female workforce. In 1891, eight thousand women worked as telephone operators; by 1946, nearly a quarter million. Women were a nimble workforce, capable of working collaboratively in networks and fluid groups—we still speak of secretarial "pools"—adaptable to the needs of the enterprise. They staffed switchboards, kept records, took dictation, and filed documents, forming, once again, the physical infrastructure of an emerging technological age. These rote office tasks are now increasingly performed electronically by digital assistants and automated telephonic systems, many of which still speak, in the default, with female voices.

As female voices buzzed across the growing telephone networks in the first half of the twentieth century, the term "girl" was used interchangeably with "computer." One member of the Applied Mathematics Panel, a division of the National Defense Research Committee that administered a human computing group in the early 1940s, ballparked a unit of "kilogirl" energy as being equivalent to roughly a thousand hours of computing labor. The National Advisory Committee for Aeronautics—the predecessor to NASA—kept its own pool of "girls," which included black women as early as the 1940s, working in a segregated west section of Langley Research Center. One of these, the mathematician Katherine Johnson, who joined the Space Task Force in 1958, hand calculated trajectories for Alan Shepard's and John Glenn's spaceflights. The Computing Group at Langley ran all its analytical calculations by hand, using the material ephemera of the gig: slide rules, magnifying glasses, curves, and early calculating machines. Johnson is often quoted as saying that she was a computer back in the days "when the computer wore a skirt."

The last significant human computing project in the United States, a reference book of mathematical tables funded by the Works Progress Administration—and overseen by another female mathematician, Gertrude Blanch—was published just as computing machines made it effectively obsolete. Human computing thrived as a stopgap between the emergence of large-scale scientific research and the capacity of hardware to carry out its calculations; eventually, the tireless machines that emerged from the spike in computer science research during the Second World War wore down their competition. After that war, the machines took over, decisively and permanently, shifting the definition of the word "computer" for the first and last time. The job description, which once required a unique cohesion of human effort, changed too: onetime human computers went from rivals to keepers, no longer executing the functions of the machine but rather *programming* those functions to be executed.

Human computing offices performed in girl-years the number crunching that machines can now perform in fractions of a second. But for a few centuries, groups of women working in hives and "harems" were the hardware: distributed biological machines capable of prodigious calculations beyond the mental capacities of any single individual, calculations that cataloged the cosmos, charted the stars, measured the world, and built the bomb. That the mathematical labor might have been, in some cases, broken down into relatively simple steps for each individual is beside the point. It's the accumulation of all those steps, executed simultaneously and collectively, that prefigured our connected, calculating, big-data world. Alone, women were the first computers; together, they formed the first information networks. The computer as we know it today is named for the people it replaced, and long before we came to understand the network as an extension of ourselves, our great-grandmothers were performing the functions that brought about its existence.

The arrival of computing machines may have emptied human computing offices, but it didn't push women from the field. Quite the opposite: many women who had been computers themselves found work tending their replacements. Female hands lifted from pencils and slide rules to desk calculators and switches, then to relays and punch card tabulators. Coaxing information into and out of the new machines was considered a woman's job, too, on the level with typing, filing documents, and patching phone calls from place to place. Not that it was easy. Dealing with early mechanical computers required a keen analytical mind and limitless patience. Just like the women whose math moved mountains, early computer programmers and operators were tasked with enormous, intractable problems. Their creative solutions often meant the difference between life and death.

Chapter Two

AMAZING GRACE

Grace Hopper was thirty-six, tenure tracked, and married when Japan attacked Pearl Harbor. She taught mathematics; her husband, Vincent, literature. The couple spent their summers fixing up an old farmhouse in New Hampshire, on sixty acres of land they'd bought during the Depression for $450. They played badminton, and Grace hooked rugs, a skill she'd picked up as a kid, summering at the family compound on Lake Wentworth.

Grace and Vincent lived the usual headaches of married academics. As Grace began her graduate studies at Yale, Vincent was working toward his doctorate at Columbia. Somehow she made time to help him research his eight-year-long thesis project, a history of number symbolism, by reading Syrian, Babylonian, and medieval texts on the subject. When she started teaching at Vassar in 1931, she audited courses in her spare time, picking up fluency in astronomy, geology, physics, and architecture. Her intellectual ambidexterity was legendary on campus: to impress students, she'd sometimes write a German sentence on the chalkboard with her left hand, and when she got to the middle, she'd switch to her right hand and finish the sentence in French.

When Grace was a junior teacher at Vassar, she picked up the classes students dreaded and nobody else wanted to teach, like calculus, trigonometry, and mechanical drawing. To revitalize them, she updated old schoolwork with new concepts, as do all good teachers. To make topography fun, she'd tell her mechanical drawing classes they were tracing the borders of fantastic imaginary worlds, and she updated the ballistics problems common to calculus textbooks to involve rockets, which were then beginning to capture the public imagination. As a result, her classes swelled with students, drawn in from departments across the college. It earned her the respect of her superiors and the unbridled resentment of her colleagues.

In the winter of 1941, Grace and Vincent were in New York City. Vincent had found a job teaching general literature at New York University's School of Commerce, and Grace had arranged a yearlong faculty fellowship from Vassar to study at NYU herself, under Richard Courant, one of the few major figures in applied mathematics. It was a nice vacation from the breakneck weekly commute they'd been driving along the Hudson, between Poughkeepsie and the city, in a Model A Ford she called Dr. Johnson. Grace liked Courant, who specialized in differential equations with finite differences, something she'd learned "one jump ahead of the students" to teach her calculus course at Vassar. Courant had a cute accent—he was a German émigré—and his lectures were always engaging. She enjoyed tackling unorthodox problems under his tutelage, even if he sometimes scolded her for taking equally unorthodox approaches to them. All in all, it was a "gorgeous year." Then everything changed.

Grace and Vincent heard the announcement on a tinny little radio, sitting at a double desk in the study they shared, surrounded by books: a violent and sudden attack at a naval base in Hawaii had left 2,403 Americans dead. The following day, the United States declared war on Japan; within a week, the conflict extended to Japan's allies, Germany and Italy.

Everybody in Grace's life wanted into the war. Vincent tried for a commission but was turned down for wearing glasses. Grace's brother, scrawny as her whole family was and with a blind spot about level with a chalkboard, didn't make the cut, either. Undeterred, they both volunteered under the draft and got in. Grace's cousin became a nurse. By the summer of 1942, everyone seemed to be gone; all the men enlisted, all the women in her family in the military's new female branches, save her sister, who had children. Grace wanted to do her part, too, but she was sixteen pounds underweight and considered too old for service. Mathematics professors, being a classified profession, weren't allowed to enlist without a release. She took a summer appointment at Barnard College to teach special war-preparedness mathematics courses for women, but it wasn't enough. All summer, midshipmen would march by the Barnard dormitories from a training ship on the Hudson, and Grace would watch them, longing to be in the navy, too.

Back upstate, she chafed with loneliness and directionless patriotism. "I was beginning to feel pretty isolated sitting up there," she said, "the comfortable college professor." She aggressively lobbied Vassar to let her go into the service. She gave the college an ultimatum, which wasn't much of one: six months or she'd leave anyway. And even though she was too old, and too thin, and her eyesight wasn't much better than her brother's, she did. The day those bombs fell on Pearl Harbor, the path of completely respectable middle-class life had been at Grace Hopper's feet, but she wouldn't take one step further. Within a few years, everything forked: she separated from Vincent, she quit her job, and she joined the U.S. Navy. It wasn't the first remarkable thing she'd ever done, and it would not be the last.

Grace turned thirty-seven on her first day at the United States Naval Reserve Midshipmen's School in Northampton, Massachusetts. She picked up the navy talk quickly—bulkheads, decks, and overheads. She'd always been good with languages. She'd taught herself German, Latin, and Greek by reading closely with a dictionary at her side, corralling the new words into each sentence like mathematical variables. Mastering military protocol was trickier, especially because it was so often at odds with social expectations. Rank and civility collided in doorways. Sometimes she'd stop to let admirals go through the doors first, but they'd try to treat her like a lady, a comedy of errors. "We usually ended up going through together," she recounted. "Which was bad." But she liked the drills. She thought they were like dancing.

She was smaller than the other recruits, and older, training alongside the students she'd been teaching only months before. But after a career in academia, commuting around the Northeast while trying to maintain two homes and a strained marriage, the constraints of military life felt like a vacation. She didn't need to think about anyone else anymore; she didn't even need to pick out her own clothes in the morning. There were few comforts—even nylons were rationed—but her domestic responsibilities had disappeared. "I just reveled in it," she told a historian years later. Unlike the youngsters she enlisted with, she "had the most complete freedom . . . I just promptly relaxed into it like a featherbed and

gained weight and had a perfectly heavenly time." With meat rationed, she ate fresh fish from the New England coast and lobster every Sunday night. She was named battalion commander and graduated first in her class, in itchy lisle stockings.

Although Grace was certain the navy would have sent her to sea had she been a man, the newly minted Lieutenant Hopper would never spend a day on board a navy ship. Instead, something in her employment history rang a bell—of all things, her study of finite differences at NYU, under Richard Courant. The navy changed Grace's orders overnight. In training, she'd assumed her military career would be spent cracking enemy codes with the elite group of mathematicians and logicians at the Communications Annex, the navy's cryptographic brain trust, overseen by one of Grace's former Yale professors. She even studied cryptography to prepare for that eventuality. Instead, the navy sent her to Harvard, where, as she liked to say, she became the third programmer of the world's first computer.

When she arrived at Harvard in July 1944, she promptly got lost. The Navy Liaison Office was nowhere to be found, and Grace hadn't been given any information about where she was to be stationed, or why. She wandered the campus, until she was finally led into the basement of the university's Cruft Physics Laboratory by an armed guard. A hawkish, six-foot-four man with an exaggerated widow's peak greeted her at the door, already irritated. The first words out of his mouth: "Where have you *been*?" Taken aback by the sight of him, she said she'd just come from Midshipmen's School and had spent the morning looking for the right place. "I was a little bewildered and at that point of course thoroughly scared of a commander," she remembered. "I told them you didn't need to do that," he muttered. He didn't think women needed service training. He asked if she'd found a place to live yet. She told him she'd only just arrived. "Well," he answered, "get to work and you can get a place to live tomorrow."

Get to work she did. Grace never saw any action during the war, but she did tame two beasts. The first was this bristly man, Lieutenant

Commander Howard Aiken. While a graduate student in physics at Harvard, Aiken—a great admirer of Charles Babbage—had designed a mechanical arithmetic device capable of solving any problem, from basic arithmetic to differential equations, that could be simplified down to numerical analysis. It was a matter of convenience: his own doctoral dissertation had been a nightmare of extensive, tedious calculations. His machine, built by IBM in exchange for the rights and donated to the university for wartime use, would be Grace's second beast. Because Aiken had imagined it as a series of daisy-chained calculators doing the work of a dozen men, it was an Automatic Sequence Controlled Calculator. Everyone at Harvard called it the Mark I computer.

The Mark I was assigned to the navy's Bureau of Ordnance to run ballistics problems for the war effort, and Aiken needed mathematicians who knew their way around differential equations with finite differences, precisely what Grace had been studying under Richard Courant that glorious year before the Japanese bombed Pearl Harbor. But Grace didn't know any of this yet. As she made Aiken's acquaintance, she heard a racket in the next room. Aiken led her to the source of the sound. "That is a computing engine," he said. Grace examined the thing, stunned. "It was all bare," she remembered; weighing in at ten thousand pounds, the Mark I stood a hulking eight feet tall, with thousands of moving parts and some 530 miles of wiring. Its inner workings were exposed, churning and noisy. "All I could do was look at it," she recalled. "I couldn't think of anything to say at that point."

The Mark I was closer to Charles Babbage's mechanical engines than to a computer in the modern sense of the word: inside its steel casing, a spinning driveshaft powered by a four-horsepower motor drove a sequence of gears and counter wheels along the entire installation. Code for the Mark I was written by hand, in pencil, on standardized code paper, and then transferred—literally punched—onto spools of three-inch-wide tape, much like the score sheet for a player piano or the pattern card of a Jacquard loom. The positions of holes in the tape, using a unique eight-bit code, corresponded to the numerals, process, and application of a given calculation. Although the Mark I was programmable in the sense that it accepted these punched-roll tapes, the distinction between hardware and software at that time was blurry, even nonexistent: every calculation called for switches to be flipped, cables to be patched.

Howard Aiken introduced Grace to her crewmates, two navy ensigns who'd arrived at Harvard while she was still in Midshipmen's School. She found out later that they'd been bribing each other to get out of sitting next to the new recruit; "they'd heard this gray-haired old schoolteacher was coming and neither one of them wanted the desk next to me." Aiken gave her a codebook, just a few pages of alien commands, and an assignment: to write a program for the Mark I that would compute the interpolation coefficients for the arctangent to an accuracy of twenty-three decimal places. "And then he gave me a week to do it in," she said, "to learn how to program the beast and to get a program running." The problem itself was not particularly mysterious for Grace—she did have a PhD in mathematics, after all. It was the machine she found inscrutable. It had no manual, and there was no precedent from which to draw, as the Mark I was the first of its kind. Grace was good at a lot of things, but she didn't have an engineering background, and she didn't know switches from relays. Aiken was testing her.

A born autodidact, she threw herself into the challenge. She pored over the codebook and picked the brains of the two ensigns, mostly the twenty-three-year-old Richard Bloch, a recent Harvard graduate and math whiz who would become her closest collaborator. Some IBM engineers were still milling around, debugging the machine; she gleaned what she could from them, too. She stayed late every night, bootlegging an engineering education by examining the Mark I's blueprints and circuit diagrams. Sometimes she slept at her desk. Years later, when Grace was an established figure in the new field of computer programming, she'd always assign the hardest jobs to the youngest and least experienced members of her team. She figured they didn't have the sense to know what was impossible.

Her first year at Harvard was nonstop, and as new programmers joined the team, Grace ascended the ranks. Using the same diligence and ingenuity she'd brought to teaching, she made herself invaluable. The gray-haired schoolteacher from upstate New York met eminent mathematicians, engineers, and pretty much everyone in the microscopic world of computing. "It was fascinating," she said, a "hotbed of ideas and concepts and dreams and everything under the sun." The computing project was in such high demand during the war that Aiken designed a second computer, the Mark II. Grace learned that one, too.

Like his machines, there was no manual for Aiken. He was temperamental, petulant, and obsessive about details. He took great pride in being the commanding officer of his own invention. Although the Mark I was built by IBM and tucked away in an Ivy League basement, Aiken ran its operation like a naval facility. Discipline was strict. His entire staff was expected to show up in full uniform and call him "commander." The computer was a "she," like any navy ship. Aiken worked people ragged; when mistakes were made, he was prone to "bawling out" the perpetrator. His criticism could be so immediate, and so fervid, that Grace often did her debugging after hours for the sake of peace and quiet. But she learned to think of her boss as a machine himself. "He's wired a certain way," she told Bloch, who was often getting into trouble with their superior. "If you understood Aiken and understood how he was wired, he was excellent to work with. I never had any difficulty. But if you tried to tell him what was right, heaven help you."

Aiken's commitment to military hierarchy was harsh, but it ultimately worked in Grace's favor: treatment in his Computation Laboratory, by and large, was commensurate to rank and ability over gender. Uniforms and formal titles helped dissolve traditional roles, as did the laboratory's complete isolation from the outside world. And although Aiken had never wanted a woman officer in his ranks, he was forced by his adherence to protocol to accept Grace's assignment. And anyway, as Grace told Howard Aiken, he was *going* to want a woman around.

She was right. She eventually became Aiken's "right-hand girl," and it wasn't long before she was solely responsible for the Mark I. She wrote the code that solved some of the war's thorniest mathematical problems, and she even wrote the missing manual for the computer, a truly laborious five-hundred-odd-page document full of circuit diagrams and operational codes. Along with her colleague Richard Bloch, she developed a system for coding and batch processing that turned the lab into the most efficient data-processing center of its day. She maintained order in a grueling wartime environment that felled lesser ensigns. And beyond her fundamental competence, there *were* some material perks to having a woman on the team. When the Mark I was having mechanical issues, Grace would sometimes "pull her mirror out of her pocketbook and stick it in front of the cams and look for sparks." By the end of his career, Aiken had but one assessment of his colleague, his highest commendation: "Grace was a good man."

Like a navy submarine, the Mark I was staffed twenty-four hours a day by a crew working in eight-hour shifts, and the computer was up and running an impressive 95 percent of the time during the war. The demand for wartime calculation was relentless, and time-sensitive requests came to the Computation Laboratory from all corners of the conflict. Grace, who had always been an omnivorous thinker, auditing courses at Vassar on every subject imaginable, took to the work. She learned to translate complex oceanography, minesweeping, proximity fuse, and ballistics problems into simple arithmetical steps, making regimented order of a messy, violent world.

The Mark I's calculations were impeded by all manner of failures: faulty code, faulty relays, and machine stoppages signaled by ominous clangs and shudders. To stay ahead, Aiken's team often worked late. One night, in September 1945, a large moth flew into the computing room through an open window, drawn in by the light on the machine. Grace found its corpse not long after, beaten senseless by one of the relays. She scotch-taped it into her log from that day, with a note: *first actual case of bug being found.* "Bug" is engineering slang that dates to at least the 1800s—even Thomas Edison used the word to refer to mechanical glitches, to "little faults and difficulties"—and Grace was known around the lab for her blackboard doodles of little bugs and monsters, each the cause of some lab snafu: a dragon who chewed holes in the punch tape, and a "gremlin that had a nose that picked up holes and put them back in the tape." After the moth incident, she bought a box of plastic bedbugs in town and scattered them around the back of the computer on a lark, causing a two-day panic.

During the war, the Computation Laboratory was isolated from the handful of other computing projects in the world, and Grace Hopper, handling the lab's everyday computational needs, had neither the time nor the opportunity to see what the rest of the field was doing. But sometimes the field came to her. Grace had been working in the Computation Laboratory for only a few months, for instance, when the physicist John von Neumann came to visit. Von Neumann had mobility; he spent much of 1944 visiting different computing projects in the United States, looking for a machine brawny enough to crack a complex partial differential equation. The Mark I was the first large-scale computer on his tour, and for three months that summer he decamped in a conference room at Harvard, outlining his problem on a blackboard while Richard Bloch set it up on the computer. Grace, still new at the lab but handy with a differential equation, assisted every step of the way.

Neither Grace nor Richard knew the specifics of the problem's application; to them, it was only an interesting mathematical challenge. And von Neumann was a character, a garrulous Hungarian theoretician who was as much of a celebrity in his day as his Princeton colleague Albert Einstein. As Bloch and von Neumann worked on the problem, they'd run back and forth between the conference room and the computer, von Neumann calling out numbers just as the Mark I would spit them out, "ninety-nine percent of the time," Grace observed admiringly, "with the greatest of accuracy—fantastic." After three months, von Neumann took their results back to a desert town in New Mexico called Los Alamos, where he was consulting on the Manhattan Project. The partial differential equation turned out to be a mathematical model for the central implosion of the atomic bomb. Grace never knew, until the bombs fell on Nagasaki and Hiroshima, precisely what she had helped to calculate.

There was not always time for Grace to consider where all the math *went,* and to what end. The calculations kept coming, some—like von Neumann's—almost inconceivably complex. To save on processing time,

Grace and Richard invented coding syntax and workarounds that set the groundwork for the way code is written to this day. As early as 1944, Grace realized she could save herself from rewriting code from scratch for each problem by holding onto reusable scraps, which came to be known as subroutines. In wartime, this was done informally: coders on the crew would share their notebooks with one another, copying over relevant bits and pieces longhand. Eventually, this practice was formalized, and future computers were built with libraries of subroutines already in place, enabling even novice coders to call on tidily packaged sequences of program instructions. When Grace's code got thorny, she made a habit of annotating the master code sheets with comments, context, and equations, making it easier for colleagues to unravel her handiwork later. This system of documentation became standard practice for programmers, and it still is: good code is always documented.

Efforts like these, which simplified and broadened the accessibility of computer programming, were Grace's calling card. Back before the war, when she was still teaching at Vassar, she'd make her students write essays about mathematical problems, because there was no sense in learning math if you couldn't communicate its value to anybody else. When she reentered the civilian world to work for the first commercial computer company, she would continue with that logic. Grace's most lasting contributions to the emerging field of computer programming all have to do with democratizing it: she pushed for programming advances that would radically change the way people talk to computers. With her help, they wouldn't need advanced mathematical terms, or even zeros and ones. All they'd need is words.

THE ENIAC SIX

The war was over before Grace had the chance to set eyes on any computing installation but the one at Harvard. And yet, there was another only three hundred miles to the south, at the University of Pennsylvania's Moore School of Electrical Engineering. Like Aiken's machines, its construction had been funded by the military to crunch numbers for the war effort. This room-sized installation of conduit and steel was the Electronic Numerical Integrator and Computer (ENIAC).

Technically, the ENIAC was faster than the machines on which Grace had cut her teeth. Where Howard Aiken's Mark I could trundle through only three calculations a second, the ENIAC was equipped to handle five thousand. This almost unbelievable warp-speed jump in processing was due to the fact that the ENIAC didn't rely on mechanical relays, gears, or driveshafts; instead, some eighteen thousand vacuum tubes, like slim light bulbs, served as its computing switches, flitting off and on in the darkness of the machine. Uncoupled from the limitations of grinding machinery, the ENIAC's vacuum-tube switches illuminated a new, ineffable realm of electronic pulses and signals. Computing would never look back.

Because these early computers were developed under wartime secrecy, computing history is full of conditional, contested touchstones—and more than a few acrimonious debates about the provenance of the "first" computer. Several machines qualify, and so the title is meted out in different ways: the Mark I, for instance, was the first *electromechanical* computer, while the ENIAC, which transcended the Mark I's physical limitations, was the first—and fastest—*electronic* computer. Across the ocean, in labs just as secret, British scientists built similar machines, each earning its own qualifier: stored-program, general-purpose, digital, binary. In these early days, every computer was an island.

When Grace Hopper visited the University of Pennsylvania in 1945, she was shocked to discover just how different the ENIAC was from the Mark I and Mark II computers with which she'd become so familiar. "The tremendous contrast," she noticed, "was the programming." Although Grace was an expert coder, she wouldn't have been able to work on the ENIAC without special training. The principles may have been similar, but the hardware and the programming approaches developed to exploit that hardware were unique. Where Grace was accustomed to writing code on paper reels, the ENIAC had to be physically reconfigured for every problem, with sections of the massive machine plugged together, essentially becoming a custom computer with every job. Any gains the ENIAC's vacuum tubes made in processing speed, the machine lost in setup time: where a calculation might take only a second to run, it could take a day to prepare, by which time the slower Mark I, with its nifty punched-roll tape, would have already been done and onto something else. A case of slow and steady winning the race—possibly for the last time in the history of technology.

Visiting Penn, Grace discovered something else that likely impressed her just as much. She wasn't the only female computer programmer on Earth. The ENIAC lab was full of women. In 1944 alone, at least fifty were working on the ENIAC in different capacities, as draftswomen, assemblers, secretaries, and technicians. Of these ranks, six women handled the time-consuming and intellectually demanding job of readying mathematical problems for the computer, plugging them in, then executing, debugging, and executing again to achieve the final results. Three among them, like Grace, were math majors. The other half had their mathematics schooling supplemented by U.S. Army training. These women, known to history as the ENIAC Six, were the peers Grace Hopper never knew she had. Some would later become her colleagues; a few, eventually, her friends. They were Kathleen "Kay" McNulty, Betty Jean Jennings, Elizabeth "Betty" Snyder, Marlyn Wescoff, Frances Bilas, and Ruth Lichterman.

The ENIAC Six were all former human computers, pulled from the Moore School's computation section, a lab that employed some one hundred mathematically inclined women. While Grace had come to Harvard straight from basic training, they'd spent the early years of the war effort in a basement, hand calculating firing tables: small, printed books shipped out with every new weapon sent to the front lines. Soldiers used these books to determine precisely at which angle to fire their guns—"basically *Angry Birds,*" as one historian of the ENIAC has pointed out—in order to hit their target. As with the arc of *Angry Birds* projectiles, external factors like weather and the drag placed on the shell by air resistance affected the point of impact, and these variables were accounted for by the female computers calculating mathematical models back on the home front.

It took a human computer about forty hours to calculate a single ballistics trajectory in all its variations. This meant that during the war, the U.S. Army was dispatching weapons faster than it could produce instructions on how to use them. The demand for human computers was ceaseless, and when the army's Ballistics Research Lab ran out of female math majors in the Philadelphia area, it began a national hiring search. One recruitment ad reached Betty Jean Jennings through a supportive calculus teacher in northwest Missouri; Ruth Lichterman caught one on a bulletin board at Hunter College in New York. Once at Penn, they joined the women scribbling the imaginary arcs of distant artillery shells for the boys on the front.

The Moore School women worked using pencil, paper, and a giant analog calculator called a differential analyzer—an onerous tabletop machine based on a design from 1920, which used gears and shafts to provide an analogy to the problem. The analyzer was fairly inaccurate, so the women interspersed its results with their own hard-won hand calculations, smoothing out the differences to create the final firing table. It was an imperfect system: arduous, fallible, and certainly far too slow for such a quickly developing, modern war. Although the computers worked six days a week, in two shifts, they could never keep up with demand. Folks around the Moore School began to entertain other possibilities.

Around this time, in 1941, the army sponsored an intensive, ten-week electrical engineering course at the University of Pennsylvania. Grace Hopper had taught a similar kind of course at Barnard before enlisting: it was tuition-free, geared toward practical applications in defense, and open to anyone with a math or engineering degree. One of the students in the Penn course was a physics professor from nearby Ursinus College named John Mauchly, a genial tinkerer who would, in his later years, sport horn-rimmed glasses and a puckish goatee. While in the class, he began to kick around the idea of a computing machine that used vacuum tubes. He discussed it with the course's laboratory instructor, J. Presper Eckert, or "Pres." Pres wasn't a star student, but he was known around the Moore School as a capable and inventive engineer. He'd shown promise early, bumming around the Philadelphia lab of Philo Farnsworth, the inventor of the television, as a kid; by the time he was an undergraduate, professors consulted him on circuit designs.

Pres thought John Mauchly's vacuum tube idea was interesting. Everybody knew vacuum tubes were too delicate for use in a computer—like light bulbs, they blow out—but Pres figured that if the tubes weren't pushed to their limit, they'd hold steady. The pair started designing circuits. John took a teaching position at Penn, happy to be closer to Pres, and, once settled, discovered the human computers at the Moore School. In their backbreaking calculations, he found the perfect application for his vacuum-circuit computer. He dictated a proposal to his secretary, Dorothy, who shunted off a memo to the university's civilian liaison with the military.

The memo was lost in the shuffle, as memos sometimes are. It wasn't until an informal conversation between the differential analyzer's maintenance man, a friend of John's from Ursinus College named Joe Chapline, and Colonel Herman Goldstine, a military liaison with the Ballistics Research Laboratory in nearby Aberdeen, that the idea came up again. When Chapline mentioned his friend John's electronic computer, Goldstine saw the potential immediately. He tried to hunt down the lost memo, to no avail. Fortunately, the secretary, Dorothy, managed to re-create it from her shorthand notes. Like most secretaries in her day, she was trained in shorthand, a type of rapid writing that looks like scribbles to the uninitiated. If it weren't for Dorothy's ability to code and decode what was at that time a largely female language, the original proposal for the electronic computer might have been lost.

The reconstituted memo was brought to army brass, who didn't need much convincing. John and Pres secured their funding in 1943 and started building the ENIAC right away. They hired engineers and former telephone company workers, who were good with relays, but most of the people who actually wired the ENIAC were women, part-time housewives with soldering irons on an assembly line. The most important hires were the human calculators, chosen from the best of the Moore School group, who would translate the ballistics computations they knew so well for the new machine. Nobody thought much of assigning women to this job. It seemed only natural that the human computers should train their own replacements. Further, the ENIAC looked like a telephone switchboard, reinforcing the assumption that its "operators" should be women, their task "more handicraft than science, more feminine than masculine, more mechanical than intellectual."

By 1944, construction on the ENIAC, then known as "Project X," took up most of the first floor of the Moore School building. One night, Pres and John conducted an after-hours demo for one of their new hires, Kay McNulty. They brought Kay and a colleague into a room where—behind a sign warning HIGH VOLTAGE, KEEP OUT—two of the ENIAC's accumulators were wired together by a long cable with a button on the end. One accumulator displayed a five. They pushed the button. The five jumped to the other accumulator, moved three places over, and transformed to a five thousand. John and Pres looked excited. Kay couldn't see why. "We were perplexed and asked, 'What's so great about that?' You used all this equipment to multiply five by one thousand," she said. "They explained that the five had been transferred from the one accumulator to the other a thousand times in an instant. We had no appreciation of what that really meant."

It meant that the ENIAC could calculate at speeds previously unimaginable, by human or machine. And although it was funded by the military to churn out firing tables as fast as the army could manufacture guns, the ENIAC was much more than a ballistics calculator. Pres and Mauchly had designed a *general-purpose* computer—think of the difference between Charles Babbage's one-note Difference Engine and the speculative Analytical Engine, which so entranced Ada Lovelace. It could perform an essentially limitless number of computational functions, as long as new programs for it were written. In its time at the Moore School, it would calculate the zero-pressure properties of diatomic gases, model airflow around supersonic projectiles, and discover numerical solutions for the refraction of shock waves. Hardware may be static, but software makes all the difference. And although it took some time to settle in, that truth came with a corollary: those who write the software make all the difference, too.

The ENIAC Six were an odd mix, thrown together by the circumstances of war. Betty Jean Jennings grew up barefoot on a teetotaling farm in Missouri, the sixth of seven children, and had never so much as visited a city before pulling into the North Philadelphia train station. Kay McNulty was Irish, her father a stonemason and ex-IRA; Ruth Lichterman, a native New Yorker from a prominent family of Jewish scholars; Betty Snyder, from Philadelphia, her father and grandfather both astronomers. Marlyn Wescoff, also a Philly native, had been hand calculating since before the war, and she was so adept that John Mauchly said she was "like an automaton." They all met for the first time on a railroad platform in Philadelphia, on their way to the Aberdeen Proving Ground, a marshy plot in Maryland the army had converted into a weapons testing facility. Bunked together, they became fast friends. Even after long days training on the IBM equipment they would be using to tabulate and sort ENIAC data, they stayed up late talking about religion, their vastly different family backgrounds, and news of the secret computer. "It was just a great romance, I think," John Mauchly hazarded when asked why these women volunteered for a job they knew so little about. "There's a chance to do something new and novel—why not?"

The reality might have been more pragmatic: in the 1940s, a woman with mathematical inclinations didn't have many options in the job market. When Kay McNulty approached her college graduation, she had a hard time finding any employment that might make use of her math major. "I don't want teaching," she explained. "Insurance companies' actuarial positions required a master's degree (and they seldom hired women, I later found out)." If the only other options are teaching math at a secondary-school level or executing tedious calculations for an insurance company, the opportunity to work in a brand-new, relatively well-paying field represented a hugely exciting change of scenery for all the women who signed up.

Computing was so new a field, in fact, that none of its qualifying attributes were yet clear. During her job interview with Herman Goldstine, Betty Jean Jennings recalled being asked what she thought of electricity. She replied that she had taken a college physics course back in Missouri and knew Ohm's law. No, no, said Goldstine, from behind the desk: Was she *afraid* of it? The job would require her to set switches and plug in cables, he explained, and he wanted to make sure she wouldn't get spooked by all the wiring. Betty Jean said she could handle it.

The ENIAC Six trained on paper, writing programs for a machine they hadn't met. When they were finally shown the finished ENIAC in December 1945, what they encountered was a massive, U-shaped assemblage of black steel housed in a room big enough to hold it along with some miscellaneous furniture. It had forty panels, grouped together to create thirty different units, each addressing some basic arithmetic function: accumulators for addition and subtraction, a multiplier, and a combination divider and square rooter. The sprawling visual effect of the machine was overwhelming. Programming, the six learned, would not be a desk job. The women would stand *inside* of the ENIAC to "plug in" each problem, stringing the units together in sequences using hundreds of cables and some three thousand switches.

There were no instructions to read, no courses to take. The only manual for the ENIAC would be written years later, long after the women had reverse engineered it from the machine itself. Built by electrical engineers, the ENIAC came with nothing but block diagrams of circuits. Just as Grace Hopper had before them, they taught themselves what to do, becoming hardware adepts in the process.

They started with the vacuum tubes and worked their way to the front panels. Betty Snyder borrowed maintenance books for the machine's punch card tabulator from a "little IBM maintenance man by the name of Smitty," who told her he wasn't allowed to lend them out but did anyway, just for a weekend, so she could figure out how the ENIAC's input and output worked. They found a sympathetic man to let them take a plugboard apart and make their own diagram for reference, even though his supervisor wasn't sure they'd be able to put it back together again (they were). It was hot and there was construction everywhere, including in the room above the one in which they worked. One day John Mauchly popped in and said, "I was just checking to see if the ceiling's falling in." They started going to him with questions, and eventually made headway.

Knowing how a machine works and knowing how to program it are not the same thing. It's something like the difference between an intellectual understanding of internal combustion and being a fighter pilot. John Mauchly and J. Presper Eckert essentially built a jet, gave the keys to six women without pilot's licenses, and asked them to win a war. It was daunting, but it presented an opportunity for the women to claim space for themselves in a field so young it didn't have a name. "At that time it was new and no one knew what to do," explained Betty Jean Jennings. Not even the men who designed the ENIAC had given much thought to how it would run. They'd ignored the actual workflow of setting up problems. In 1973, Mauchly himself admitted that he and Pres had been "a little cavalier" about programming, saying that they

Betty Jean Jennings (left) and Frances Bilas (right) operate the ENIAC's main control panels.

"felt that if we had the machine capable . . . there would be time enough to worry about those things later."

As it turned out, Mauchly found other people to worry about those things—six people, in fact, in wool skirts and thrilled by the challenge. "How do you write down a program? How do you program? How do you visualize it? How do you get it on the machine? How do you do all these things?" wondered Betty Jean. It would be up to the ENIAC Six to figure it out.

Today, programming can be tricky, but it's accessible. To write code, you don't need to study circuit diagrams, take apart components, and invent strategies from scratch. Instead, you simply need to learn a programming language, which acts as an intermediary between coder and machine, just as a shared spoken language can bridge a gulf of understanding between people. You tell the machine what to do in a language you both understand; the machine then translates and executes your commands on its own. The ENIAC had no such language. The computer accepted input in only the most elemental of ways, and so the ENIAC Six rolled up their sleeves and met the machine on its level. As Betty Jean Jennings recounted:

> Occasionally, the six of us programmers got together to discuss how we thought the machine worked. If this sounds haphazard, it was. The biggest advantage of learning the ENIAC from the diagrams was that we began to understand what it could do and what it could not do. As a result we could diagnose troubles down to the individual vacuum tube. Since we knew both the application and the machine, we learned to diagnose troubles as well as, if not better than, the engineer.

Unlike Grace Hopper, who managed a team of operators punching her handwritten code into the Mark I's tape loops, the ENIAC Six moved around inside the great machine itself. They replaced individual burned-out vacuum tubes from among thousands—several burned out every hour, despite Pres's design—fixed shorted connections, and wired control boards. They wrote programs, feeding them gently into the machine with much trial and error. The job required a combination of mechanical dexterity and mathematical know-how, to say nothing of organizational skills: punched cards containing the ENIAC's programs needed to be sorted, collated, tabulated, and printed. The word "programmer" didn't exist yet, but Betty Snyder thought of herself as a "cross between an architect and a construction engineer." Betty Jean Jennings was more blunt. "It was a son of a bitch to program," she wrote.

Unfortunately, none of this effort did the U.S. Army any good. Although it ran a number of one-off calculations, the war ended before the ENIAC became fully operational as a ballistics calculator. In peacetime, however, the ENIAC was no longer secret, and the computer was unveiled to the public in 1946, with much fanfare and two different demonstrations. The first, for the press, was by all accounts a bit lackluster. The second, for the scientific and military community, was a hit, thanks largely to a demonstration of a trajectory calculation programmed by Betty Jean Jennings and Betty Snyder.

The two Bettys, as they were sometimes known, were the aces of the ENIAC programming team; after the war, they both went on to long and pioneering careers in the commercial computer industry. As was common in the history of human computing, the pedagogy of the Moore School emphasized working partnerships, with teams of two people seeking out errors in each other's work. Betty Jean and Betty were ideal partners, because they delighted in finding each other's mistakes. They both wanted perfect code and never let their egos get in the way of achieving it. "Betty and I had a grand time," Betty Jean wrote in a memoir. "We were not only partners, but we were friends and spent as much of our free time together as possible."

A few days after the first ENIAC demonstration, Herman Goldstine, their military liason, and his wife, Adele, invited Betty and Betty Jean over to their apartment in West Philadelphia. Adele trained the human calculators at Penn and had always struck Betty Jean as an impressive, big-city woman; at the Moore School, Adele lectured sitting on her desk, with a cigarette dangling from the corner of her mouth. Betty Jean was surprised to find the Goldstine apartment rather ordinary, with few personal touches and a set of twin beds. As Adele served the Bettys tea and the Goldstine cat leaped uninvited onto their laps, Herman asked them if they could set up a ballistics calculation on the ENIAC in time for its unveiling to the scientific community twelve days later. It was a big ask, and Betty Jean sensed that Herman Goldstine was nervous about the demonstration. Well-known scientists, dignitaries, and military brass would be there, and everyone was keen to see that the ENIAC worked as advertised. Not much has changed, it seems, about the way tech keynotes are anticipated and prepared.

The Bettys asserted vigorously that yes, absolutely, they could make it happen. They were bluffing. Although they'd spent the last four months working out a ballistics trajectory program on paper, they hadn't actually plugged it into the ENIAC yet, and they had no idea how much time the transfer would take. They started the next day.

Betty Snyder was twenty-eight; Betty Jean Jennings had only just turned twenty-one. They knew they'd been asked to do something important and that everyone they worked with was counting on them. The pair worked around the clock for two weeks, living and breathing the trajectory program. Their colleagues Ruth Lichterman and Marlyn Wescoff supported them by hand calculating an identical trajectory problem on paper, mirroring step-by-step how the ENIAC would process the calculation. This would help the Bettys debug the ENIAC if it made any errors. Men popped by with offerings: the dean of the Moore School left them some scotch, and John Mauchly came in on a Sunday with a bottle of apricot brandy. They didn't really drink—maybe a Tom Collins on special occasions—but Betty Jean kept a taste for apricot brandy for the rest of her life.

The night before the big demonstration was Valentine's Day, but the Bettys didn't go on any dates. Their ENIAC program had a massive bug: although they'd managed to model the trajectory of the artillery shell perfectly, they couldn't figure out how to make it *stop*. When their imaginary shell hit the ground, the mathematical model kept going, driving it through the earth with the same velocity and speed as it had while shooting through the air. This made the calculation worse than useless. If they didn't find some way to stop the bullet, they'd embarrass themselves in front of eminent mathematicians, the army, and their employers. In desperation, they checked and rechecked settings, comparing their program with Ruth and Marlyn's test program, but they were stuck. A little before midnight, they left the lab. Betty took the train home from the university campus to her house in suburban Narberth. Betty Jean walked home in the dark. Their spirits were low.

But Betty Snyder had one trick left: when stuck on a logical problem, she always slept on it. Wearily, she spent her hour-long train ride home that night considering the artillery problem and its various potential solutions. When she fell into bed, her subconscious began to untangle the knot. The next morning—February 15, 1946—Betty arrived at the lab early and made a beeline to the ENIAC. She'd *dreamed* the answer, and knew precisely which switch out of three thousand to reset, and which of the ten possible positions it should take. She flipped the switch over one position, solving the problem instantly. Betty could "do more logical reasoning while she was asleep than most people can do awake," marveled Betty Jean.

The ballistics trajectory demonstration was a huge success, thanks to the Bettys' clever ballistics program and a little old-fashioned razzle-dazzle from John and Pres, who placed halved Ping-Pong balls over the ENIAC's neon indicators. During the demonstration, staff dimmed the lights in the room, showcasing the ENIAC's thinking face in feverishly blinking orbs of light. The program was faster than a speeding bullet, literally: the ENIAC calculated the trajectory in twenty seconds, faster than it would have taken a real shell to trace it. The Bettys and Kay McNulty hustled over to the tabulator, made printouts, and handed them out to the audience as souvenirs.

The event made headlines. The women were photographed alongside their male colleagues—they remember flashbulbs—but the photos published in newspapers showed only men in suits and military decorations posing with the famous machine. The press had a field day with the ENIAC, presenting it as a fruit of the war effort unveiled for the better living of the American people. Because of their unfamiliarity with computing, journalists called the ENIAC a "giant brain" and a "thinking machine," a mischaracterization that has persisted in the popular consciousness, enthusiastically supported by science fiction writers, ever since. The ENIAC couldn't think. It could multiply, add, divide, and subtract thousands of times per second, but it couldn't reason. It was not a giant brain. If there were giant brains in the room, they belonged to the people who built—and ran—the machine.

It irked the ENIAC women to read newspaper articles claiming the machine itself was clever; they knew better than anyone that it was just a room full of steel and wire. "The amount of work that had to be done before you could ever get to a machine that was really doing any thinking to me just staggered the mind," complained Betty Jean, and "I found this very annoying." It was more than annoying; it effectively erased her. The 1946 *New York Times* story about the ENIAC demonstration breathlessly reported that "the ENIAC was . . . told to solve a difficult problem that would have required several weeks' work by a trained man. The ENIAC did it in exactly fifteen seconds."

As historian Jennifer S. Light points out, that claim ignores two essential factors: first, that the "several weeks' work" would never have been done by a man in the first place. It would have been done by a female computer working long hours at the Moore School. Second, the claim that the ENIAC solved the problem in "exactly fifteen seconds" completely disregards, through ignorance or willful dismissal, the weeks of work, again conducted by women, that went into programming the problem before it was even put on the computer. As far as the press was concerned, nothing outside of those fifteen magical seconds—not the hours of coding and debugging, not the labor of programmers and maintenance workers and operators—counted. Light writes, "The press conference and follow-up coverage rendered invisible both the skilled labor required to set up the demonstration and the gender of the skilled women who did it."

After the ENIAC demonstration, once the glad-handing and photo-ops were over, the university hosted a big celebration dinner. Judging from a menu for the event, no expense was spared. Military brass and members of the scientific community ate lobster bisque, filet mignon, and "fancy cakes." None of the ENIAC Six were invited—not even the Bettys, who had created the demonstration the dinner was held to celebrate. They'd helped introduce the century to the machine that would come to define it, and nobody congratulated them. The Goldstines snubbed them completely. Even their supporters, John Mauchly and J. Presper Eckert, were too caught up in the excitement of the day to comment on the demonstration program. On February 15, just as they had late the night before, Betty Jean Jennings and Betty Snyder went home dejected. It was cold, and they were exhausted. "It felt like history had been made that day," Betty Jean wrote in her autobiography decades later, "and then it had run over us and left us in its tracks."

History would run them over again and again. Neither Betty nor Betty Jean would be credited for writing the ENIAC demonstration program until they began to tell their own stories over fifty years later. Herman Goldstine, in his influential history of the ENIAC, wrote that *he and Adele* had programmed the February 15 demonstration themselves, a move of revisionist history that Betty Jean Jennings, late in life, called a "boldface lie." In subsequent retellings, the women were skipped over repeatedly. In some historical images, the ENIAC Six are captioned as models, if pictured at all. "I wasn't photogenic," said Betty Snyder. "I wasn't included on any of the pictures of the entire stupid thing." When the army used a War Department publicity shot of the ENIAC for a recruitment ad, they cropped out the three women in the picture entirely. The War Department's own press releases about the ENIAC cited a vague, genderless "group of experts" responsible for the machine's operation, and mention by name only John Mauchly, J. Presper Eckert, and Herman Goldstine.

It's tempting to look at the historical association between women and software and assume some inherent affinity: that women appreciate the mutable, language-oriented aspects of programming, while men are drawn to the practical, hands-on nature of hardware. Some might posit as much from the Babbage-Lovelace partnership, Howard Aiken and Grace Hopper's testy relationship a generation later, or from the gendered division of labor between male hardware engineers and female operators on the ENIAC. But in all of these instances, women have fallen on the software side not because the work was somehow more suited to them but because software, still inextricably entangled with hardware, wasn't yet a category with its own value. As far as anyone understood it, software—the writing of code, the patching of cables—was really just the manipulation of hardware, and the title of "programmer" wasn't yet distinct from the more menial "operator," a rote job that leaned female because of a long female history of secretarial work. Further, the hiring of women to run computers like the ENIAC reflected a long tradition of women as computers themselves, laboring over applied mathematics in university and research settings. Women had been doing the math for as long as anyone could remember.

"If the ENIAC's administrators had known how crucial programming would be to the functioning of the electronic computer and how complex it would prove to be," Betty Jean Jennings eventually determined, "they might have been more hesitant to give such an important role to women." At the time, it was difficult to perceive of programming as an occupation distinct from simply *operating* a computer and, indeed, the ENIAC women's jobs were officially classified as "subprofessional, a kind of clerical work." It would be years before those who approached computers began to define themselves in relation to them, as programmers or computer scientists, rather than as operators or electrical engineers. It would take even longer before a vision of programming as an art form with the potential to reshape the modern world came into focus.

Chapter Three

THE SALAD DAYS

World War II was a war of technology. The relationships between government, private industry, and academia forged to gain a national edge on the Axis gave us the military-industrial complex and bankrolled a generation of technological innovators in the process. It's unlikely computing would have developed as a field or an industry as quickly as it did without the complex calculations demanded by the war machine. The war made risks worth taking and accelerated everything. It also let women into the process. Men may have dropped bombs, but it was women who told them where to do it.

It's unsettling to think of any war, especially such an ugly one, as an opportunity. But working on military calculations during World War II allowed Betty Jean Jennings, Betty Snyder, Grace Hopper, and their peers to do more with their lives than teach, marry, or be secretaries. It opened an entirely new technical field to women, one whose importance would become evident only after they showed what remarkable things could be done at the confluence of people and computing machines. But change is never so simple. As easily as war gave these women a ticket out of potentially desultory marriages and dead-end secretarial careers, peace threatened to take it all away.

After the war, as military funding dried up and authority over computational projects transitioned back to civilian hands, Grace Hopper found herself at a crossroads. In a short time, she'd become an expert in a nascent field, but she'd made sacrifices. Although she and Vincent had been separated since the beginning of the war, they only divorced in 1945. He promptly remarried—to a friend of Grace's, who had been a bridesmaid at their wedding. It can't have been easy. She was forty-three, and after what she'd been through, she couldn't dream of going back to teaching calculus to undergraduates and being the comfortable small-town college professor once again. But as soldiers coming home from the front reclaimed their place in American life, many women around Grace were returning to their prewar roles. Even a few of the ENIAC Six went domestic. Being someone who'd left everything behind to stake a career at the doubly male-dominated intersection of military and academic life, Grace was nervous.

Howard Aiken parlayed the Computation Laboratory's wartime computing work into a longer appointment at Harvard, and he kept Grace, who was indispensable to the operation of the lab, on as a three-year research associate. In her new role, Grace no longer did any programming. Instead, she assisted Aiken and wrote a user's manual for the Mark II, for which she—as with the first manual—received no byline. She led tours of the facility, explaining to distinguished visitors the many applications of the once-secret computing machine. Aiken tried to maintain the Computation Laboratory's sense of urgency, but things weren't the same; wartime atmosphere in a time of hard-won peace lent a lightly sadistic air to the everyday. Grace had always been a social drinker, but watching her role dwindle, she drank on the job, keeping a flask in her desk. In 1949, Harvard's contract with the navy expired. Aiken, safely tenure-tracked, stayed on, but the university didn't offer promotions to women, so Grace—by an order of magnitude the preeminent national expert on computer programming—lost her job in the lab she'd managed for years. "My time was up," she said.

The ENIAC lab at Penn was gone, too. The computer itself was moved to the Aberdeen Proving Ground—they'd had to tear out an entire wall of the Moore School building to remove it—and eventually seven more women would work on the ENIAC at Aberdeen, where it ran until being decommissioned in 1955. The ENIAC Six, having moved on to the next phases of their lives, would never program their namesake again.

During the war, it hadn't mattered who owned the ENIAC. It was built with the army's dollar, and it worked in the army's service. With the conflict over, however, the computer became contestable. Eckert and Mauchly wanted to patent their invention, but the university demanded all manner of licenses and sublicenses for their design, as well as rights to whatever machine they planned to build next. Penn offered them tenure in exchange for a patent release, but the two engineers didn't want to be beholden to the university for the rest of their careers. Pres and John chose to write off the ENIAC, cut their losses, and strike out on their own.

By 1947, they'd founded their own company, the Electronic Control Company, later renamed the Eckert-Mauchly Computer Corporation. Their offices on Ridge Avenue in Philadelphia were nothing fancy: an old hosiery mill with no air-conditioning and a few rows of wooden desks someone picked up at a secondhand store. Right across the street was a junkyard. On the other side of the building was the Mount Vernon Cemetery. If the computer didn't run, Grace Hopper joked, they'd throw it out one window into the junkyard and jump out the other.

The narrative of enterprising young technology innovators striking out on their own to take risks and change the world is familiar to us now. Every garage-to-riches Silicon Valley story seems to follow the same approximate arc. But the Eckert-Mauchly Computer Corporation was the first commercial computer company in the world. Until EMCC, computers had been one-off machines custom built for the calculations of war: ballistics, code-breaking, and the fluid dynamics of nuclear bombs. But computers had a huge prospective market in academia and in calculation-intensive industries like aviation, and they could make quick work of accounting and payroll problems for large corporations. Pres and John weren't the only people to realize the massive commercial potential of computers as business machines, but they were among the first to give the standardized production of computers a try. As architects of the now-world-famous ENIAC, they had a running start.

EMCC scored its key contracts early: the Census Bureau, the National Bureau of Standards, Northrop Aviation, and the Army Map Service all wanted their own version of the ENIAC. Each installation would have to be customized to the client's specific computing needs, bundled with systems, services, and support. Software wasn't an off-the-shelf product yet—or a word, for that matter—so the Eckert-Mauchly Computer Corporation would need programmers. Fortunately, the best in the world were their former colleagues at the Moore School—Betty Snyder, Betty Jean Jennings, and Kathleen "Kay" McNulty, and the grand dame of code herself, Grace Hopper, who was looking for a path forward after losing her place at Harvard. Pres and John were smart enough to hire them all.

For its first few years, the Eckert-Mauchly Computer Company was an anachronism. It was a tech startup among corporate giants, building the most advanced computers in the world, all of which ran custom code written by the talented female programmers leading its software development teams. Betty Jean Jennings remembered it as a magical time. "I loved it," she wrote, "and I never again felt so alive. We worked constantly. We arrived early. We arranged meetings during coffee breaks and lunch. We worked late." There were no formal titles, and departments were porous. People worked directly with one another when projects called for it; there was no chain of command, no bureaucracy. Everyone tackled the problems in front of them, taking initiative wherever it led. "The fact is," Betty Snyder said, "we all believed so much in what we were doing that we worked together, that's all."

The founders were widely liked and admired by their female staff, because they were good teachers, they listened, and they had vision. Meeting with clients, John Mauchly could rattle off new applications for their computers on the spot. Grace was motivated to join EMCC because she knew she'd learn something from John, whom she remembered as "a very delightful person to be with and fun to work with. He was just as excited about it as you were and he was right in the thick of it and thoroughly enjoying it and what I'd call a real good boss." Pres, too, was a "real good boss," albeit an odd one. While deep in conversation, he'd often lose track of himself, wandering along hallways, down into the basement, and back upstairs. He'd sit on tables. Once, he climbed up onto a file cabinet while talking to someone without even noticing. In all the years she worked at EMCC, Betty Jean never knew where Pres's office was: she never saw him in it. He was always among them, moving from group to group with questions and solutions. The women joked that Pres liked talking so much that he'd stop the janitor from mopping just to have someone to bounce ideas around with. Despite these idiosyncrasies, he was theirs. "We all accepted Pres as he was," Betty Jean recalled.

During a time when many companies were firing competent women to make room for returning GIs, Eckert-Mauchly Computer Company not only employed women but also valued their contributions, giving them even more significant programming responsibilities than they'd had during the war. Both Betty Snyder, by then Betty Holberton, and Betty Jean Jennings managed major projects: Betty wrote the instruction code and Betty Jean did the logical design for the company's marquee product, the Universal Automatic Computer, or UNIVAC. The early years were lean—Betty, along with many engineers, worked for nothing at first—but everyone was treated well once the company got its legs. A year into her employment at EMCC, Betty Jean discovered that a recent male hire was making four hundred dollars more than her a year, and when she took her grievance to John Mauchly, he heard her out and immediately raised her salary to two hundred dollars *above* the man. EMCC even held company-wide programming courses every spring, in the hopes of training fresh blood from elsewhere in the ranks. "That's how so many secretaries got to be programmers before we were through," Grace Hopper recalled. That was the company ethos. "A gal who was a good secretary was bound to become a programmer, meticulous, careful about getting things right. Step-by-step attitude. The things that made them good secretaries were the very things that made them good programmers."

Grace had been struggling with alcoholism during her final years in

Howard Aiken's thankless Computation Laboratory, and her friends hoped that the job at EMCC would help her stop drinking and regain her confidence. Even if she wasn't valued at Harvard, her talents were in high demand in the burgeoning computer industry: in addition to the offer from Eckert and Mauchly, in 1949 she considered a job in ballistics research at Aberdeen, making cryptological machines for Engineering Research Associates, and with the Office of Naval Research. But only EMCC had working machines available to program within the year, and Grace didn't want to wait any longer. At EMCC, she could return right away to her true passion: writing code, and surrounded by brilliant women to boot. She was grateful. Much as she'd reveled in the transition from civilian to military life back in Midshipmen's School, she "slipped into UNIVAC like duck soup," as though she'd "acquired all of the freedom and all of the pleasures of the world."

The woman Grace worked most closely with at EMCC was undoubtedly Betty Holberton, the ENIAC whiz who solved programming problems in her sleep. This was only one aspect of Betty's unique mind. From a young age, she'd been keen on puzzles. At fifteen, when her family's calculating machine was on the fritz, she took the whole thing apart, labeled every component, and laid them all on the table, precisely diagrammed. She liked to think of machines that way: as a concert of small pieces working together in a logical whole, and found that she thought "like a radar screen," going "around and around the problem" until she had it all in mind. After these radar sweeps, she was able to see errors almost instantly. Her talent was so potent that she was often brought in, "if the programmer and the engineer couldn't resolve what the problem was," to debug other people's projects. Later in her career, she'd insist on not being called in until *after* the engineers had been stuck for at least four hours. Otherwise, it was a waste of her time.

Betty was EMCC's secret weapon, and by the time Grace Hopper arrived in Philadelphia, Betty had already been with the company for two years, working one-on-one with John Mauchly, at his living room table before they rented the warehouse on Ridge Street, developing C-10, the operational code for the UNIVAC. Although she claimed not to have been "photogenic" enough to be pictured with the ENIAC, there was no denying the singular beauty of her code. C-10 was the elemental instruction set for the computer, and Betty was extremely thoughtful about writing it. Instead of relying only on numbers and symbols, she included letters as shorthand for commonly used operations: *s* for "subtract," *a* for "addition." This was a first, and it made the UNIVAC more user-friendly to operators without a mathematical background. Grace thought the C-10 code was brilliant—beautiful, even. She loved anything that made programming more approachable. And she thought Betty was the best programmer she'd ever met. Betty was more modest. According to their colleague Betty Jean Jennings—who didn't always get along with Grace—Betty Holberton would say of Grace that "if anyone could do something she couldn't do, that person obviously had to be the best in the world at it, because only the best could outdo Grace!"

Betty taught Grace how to use flowcharts to set up complex problems, something she had never done. The Mark I had been perfectly linear, all the code just "riding" along on trails of paper tape. But the machines at Eckert-Mauchly could modify instructions along the way, making loops. "Now I had two-dimensional programs to think about," recalled Grace, which required "thinking in another dimension," something Betty formalized in her elegant flowcharts. Betty, Grace concluded, "was terrific." Beyond the C-10 code, Betty made two more significant contributions to computing while she was working at EMCC. The first was cosmetic: she persuaded the engineers to change the UNIVAC's exterior color from black to the oatmeal beige that would become the universal color of desktop computers (yes, I have Betty to thank for my dun-colored Dell). The second was a program she called a "Sort-Merge Generator," which took specifications for data files and automatically generated routines for sorting and merging that data, keeping track of all the input and output in the UNIVAC's tape units.

Grace Hopper was floored by Betty's Sort-Merge Generator. According to Grace, it marked the first time a computer was ever used to *write a program that wrote a program*. This would have a huge influence on Grace, and indeed on the entire history of computing. But not everyone saw it that way at first. "At that time the Establishment promptly told us—at least they told me quite frequently—that a computer could not write a program," Grace remembered. "It was totally impossible; that all that computers could do was arithmetic, and that it couldn't write programs; that it had none of the imagination and dexterity of a human being. I kept trying to explain that we were wrapping up the human being's dexterity in the program."

Indeed, thanks to the dexterity of programmers like Betty and Grace, UNIVAC was the most powerful computer in the world. Its magnetic-tape programs allowed input and output speeds to finally match the speed of its electronic components. The UNIVAC dominated its competition, becoming synonymous with the technology itself: as Kleenex is to tissues and Xerox is to photocopies, in the 1950s, UNIVAC was to computers. Even a young Walter Cronkite consulted it like an oracle. On Election Night, 1952, a UNIVAC running a statistical program live on CBS Election Night coverage predicted Dwight Eisenhower's landslide victory over Adlai Stevenson when no traditional poll saw it coming. The gimmick was so popular that the computer became a regular feature on CBS.

Despite their public successes, Mauchly and Eckert stumbled behind the scenes. They were academics, with good intentions inversely proportionate to their business expertise. Their stored-program electronic computers were an untested, expensive technology that required a hard sell and no shortage of hand-holding for those clients who did buy in. EMCC's first major injection of capital—from a Baltimore company, American Totalisator, that made betting-odds calculators for racetracks—fell to pieces after the vice president of the company died in a freak plane crash. Strapped for cash, Mauchly and Eckert tried to sell EMCC to IBM, but Big Blue wanted nothing to do with them—IBM's aged CEO, Thomas Watson Jr., didn't believe in magnetic tape and elected to stick with the proven business of punch card systems. They were forced to sell EMCC to a business machine company, Remington Rand, once the proud manufacturer of the world's two most important weapons: guns and typewriters.

Selling out to Remington Rand wasn't an inspired decision, but it was grimly practical. The company had been in the punch card business for years, and they anticipated growth in the computing sector. They bought out EMCC's debts, effectively saving the company from ruin, and promised that their experienced sales teams could sell the UNIVAC in their sleep. Unfortunately, Remington Rand's senior management knew nothing about Boolean algebra, subroutines, or code. Worse, they had no clue what to do with all the women working in senior positions at EMCC.

"When Remington Rand bought UNIVAC," Betty Snyder recalled some years later, "as far as I was concerned, that was the end of the line, because their idea of women was to sit beside a typewriter." Nobody at Remington Rand had ever worked with technical women, and they resisted the idea that a woman could lead a presentation, meet with clients, or even understand the UNIVAC installations they had designed. "I mean, it was just as though they never dreamed that anybody would listen to you and take you seriously," Betty Jean Jennings complained. It would be the end of the salad days.

Remington Rand had the foresight to buy the Eckert-Mauchly Computer Corporation, but they didn't know what to do with the group of freethinkers they inherited in the acquisition. It was a crisis of values that would be repeated ad infinitum in the technology business. Eckert-Mauchly was a place where brilliant women coaxed symbols into code and new ideas unraveled on every spool of magnetic tape. But that vision, and the unorthodox environment that nurtured it, struggled to sustain itself in a world of punch card tabulators. The dreamers were bought by a company who "thought these idiots down in Philadelphia were insane," that "nobody would buy these million-dollar machines, and there were very few applications for which they were really usable."

The open and transparent company structure of EMCC was gobbled up by Remington Rand's old-school corporate hierarchy. The company's central office was in New York City, which may as well have been Mars for all the EMCC team was able to communicate with it. Betty Snyder, who'd never had a title or a department, suddenly had a boss.

"That was a disaster," she said, "because I'd go to him with decisions and he'd make the wrong decisions and I had to live with them." At EMCC, everyone had served the machine. But at Remington Rand, there were other masters, whose interests were often at odds with their work. In 1990, long after they'd all retired, a group of former Eckert-Mauchly employees got together at the Smithsonian Institution to get their history down on record. Even forty years later, nobody had a nice thing to say about Remington Rand. An excerpt from that oral history:

> **CERUZZI (moderator):** I wonder if there is anyone here who would like to defend Remington Rand management, because . . . [laughter] Do we have any volunteers for this?
>
> **TONIK (former UNIVAC programmer):** No, not exactly.
>
> **CERUZZI:** No? Not at all.

The UNIVAC was a million-dollar machine, but Remington Rand's salesmen were not trained to understand what it could do. They were used to tabulating machines, which ran on punched cards; hard copies were an easy sell for office contracts, as opposed to the UNIVAC's black magnetic tape, which was unknowable, literally opaque. "There was no feeling" about the UNIVAC, Betty recalled, "except what they were going to get out of it." The best the salesmen could do was use the computer in a bait-and-switch routine: they'd bring Betty Jean Jennings to sales calls at government agencies, where she would diligently describe everything the stored-program computer was capable of doing. Then they'd hustle her out of the room so that the men could sell some typewriters.

It might've been better than the alternative. When Remington Rand's salesmen *did* take a crack at selling the UNIVAC, their lack of understanding horrified the old Eckert-Mauchly programmers. At one point, Rem-Rand hired a man unknown to anyone at Eckert-Mauchly to write all of the UNIVAC marketing materials and to finally train sales staff on the machine. He consulted nobody familiar with the UNIVAC.

Mauchly was furious. "We are at a loss to understand how anyone, having accepted a position demanding experience he does not have," he banged out in a memo, "would fail to seek all possible help to discharge his duties."

Grace Hopper and her programming team were forced to pick up the slack. By 1950, they were pulling multiple shifts: selling and marketing the UNIVAC where Remington Rand couldn't do it effectively and managing customer support when the client installations inevitably posed problems. They were bound to fail, even if they succeeded: Remington Rand management forcibly scaled down production of the UNIVAC to six a year, but after two years of the programmers handling sales, they'd sold forty-two machines. "The result was they couldn't deliver the damn things," a UNIVAC engineer, Lou Wilson, recalled at the Smithsonian meeting.

Grace, in particular, found herself in an untenable position. She was doing the work of three people in a field most had trouble even defining. At Eckert-Mauchly, she'd been senior mathematician—"it sounded impressive enough to match the salary"—but at Remington Rand, seniority just meant more work. Beyond managing a team of programmers and overseeing custom software projects for each client, she served as clearinghouse for customer support. It was a constant battle. To say nothing of her side hustles: she was always furthering the art, working in what little spare time she had on improvements to programming technique. But it suited her to be busy. She'd had a rough patch in the years between Eckert-Mauchly's insolvency and its acquisition by Remington Rand, her alcoholism creeping back—she was even arrested for drunk and disorderly conduct in November 1949. But Howard Aiken's lab at Harvard hadn't been a pleasure cruise, either, and those punishing circumstances had brought out her best ideas. The same would happen during her most trying years at Remington Rand.

Chapter Four

TOWER OF BABEL

In the early 1950s, programming wasn't well understood beyond its industry—or even within it, judging from the difficulty Grace had communicating with the sales department of her own company. Not coming from any existing art, programmers began their careers elsewhere: some, like Grace, were mathematicians while others were discovered through aptitude tests or were given a shot because of their affinities for crossword puzzles and plane geometry. To succeed, they had to learn their hardware exhaustively and fail constantly. This created a certain sense of earned privilege. John Backus, a computer scientist at IBM and a contemporary of Grace Hopper's, famously characterized programmers in the 1950s as a priesthood, "guarding skills and mysteries far too complex for ordinary mortals." As much as the wizards appreciated shortcuts for their drudgery, "they regarded with hostility and derision more ambitious plans to make programming accessible to a larger population."

Grace wanted out of the priesthood. She strongly believed that computer programming should be widely known and available to nonexperts. If computers weren't so painstaking to program, perhaps they'd be easier to sell; if clients could write and rewrite their *own* code, then her staff wouldn't have to create custom programs for each UNIVAC installation. Which would be ideal, because there were only a few really good programmers in the world, and their talents were being wasted on dreck. As the world awoke to its possibilities, the computing industry was exploding, and there weren't enough trained programmers to satisfy demand. Grace and her peers weren't getting any younger, either; the field needed new blood and it needed to become more accessible. Grace knew that would only happen when two things occurred:

1. Users could command their own computers in natural language.
2. That language was machine independent.

That is to say, when a piece of software could be understood by a programmer as well as by its users, and when the same piece of software could run on a UNIVAC as easily as on an IBM machine, code could begin to bend to the wills of the world. Grace called this general idea "automatic programming," and to anyone who knew her, it was a logical outgrowth of her Harvard work on subroutines and code documentation. She'd always liked finding ways to make programming a little easier, a little more efficient. During the war, her shortcuts saved time and lives. After the war, they saved money and heartache.

To the higher-ups at Remington Rand, however, investing in automatic programming seemed like a distraction from the real business of computer sales. Why make things easier, when what they were selling was expertise? Her fellow programmers had reservations, too: Grace's proposition might put them out of work. In later years, as the discussion over automatic programming grew more and more contentious, each side of the argument earned a nickname. Those who resisted automatic programming became known as the "Neanderthals." They might as well have called themselves framebreakers, as Lord Byron had over a century before.

The "space cadets," on the other hand, believed in a bug-free future, where the programs wrote themselves, or at least pulled their own weight. As utopian as that might have seemed to the Neanderthals, who believed that programming was an expert craft inimitable by machine, the industry-wide need for automatic programming only grew. Remington Rand hadn't anticipated how much support UNIVAC required; it became a strain on the company just to keep its clients' computers operational. Grace, who was great at pleading her case to bureaucrats, made it clear that the costs associated with programming threatened to approach the cost of hardware. Remington Rand caved. They created an Automatic Programming Department and put Grace in charge. Her first order of business was to write a compiler, a kind of mediator program that simplifies writing code at the machine level. Just as mechanical computers replaced a generation of human calculators, making programmers of mathematicians like Grace, her compiler—the first of many intermediaries between people and machines—would once again redefine the nature of the job. The mathematician who had become a programmer would soon become, in turn, a linguist.

A quick lesson: computers do not understand English, French, Mandarin Chinese, or any human language. Only machine code, usually binary, can command a computer, at its most elemental level, to pulse electricity through its interconnected logic gates. Broken down, every program is just a maddeningly explicit list of instructions, in this machine code, about where and how electricity should move. The most basic programs specify operations at the hardware level—just one step above physically plugging a computer like the ENIAC together—and more complex programs are aggregations of these basic operations. In Grace's day, debugging required thinking like a machine, bit by painstaking bit. If the priesthood guarded one mystery above all, it was the secret to achieving limitless patience.

Compilers are fundamental to modern computing. They make programming languages, with their ever-higher levels of symbolic notation, comprehensible to the binary lizard brain of the computer. It's now a given that using a computer—and even programming one—requires no specific knowledge of its hardware. I don't speak binary, but through the dozens of software interpreters working in concert whenever I make contact with my computer, we understand each other. Machine code is now so distant from most users' experience that the computer scientist and writer Douglas Hofstadter has compared examining machine code to "looking at a DNA molecule atom by atom."

Grace Hopper finished the first compiler, A-0, in the winter of 1951, during the peak of her personnel crisis at Remington Rand. The following May, she presented a paper on the subject, "The Education of a Computer," at a meeting of the Association for Computing Machinery in Pittsburgh. In the paper, she explained something counterintuitive: that adding this extra step, a layer between the programmer and the computer, would increase efficiency. She used a personal example to illustrate. In the past, mathematicians like her had been stuck with the chore of arithmetic—all those tedious little steps on the way to interesting solutions. Ostensibly, a computer like the UNIVAC took over those arithmetic tasks, freeing the mathematician to think more stimulating thoughts. The reality, however, was that the mathematician became a programmer instead, again consumed by tedious little steps. Grace loved coding, but she admitted that "the novelty of inventing programs wears off and degenerates into the dull labor of writing and checking programs. This duty now looms as an imposition on the human brain."

Her solution was to insert a third level of operation, empowering the computer to write its own programs. It would do that by *compiling* selected subroutines—reusable scraps of code saved to the computer's memory—from the computer's baked-in subroutine catalog. The compiler could then automatically arrange subroutines and translate them into nitty-gritty machine code. That way, "the programmer may return to being a mathematician," and the computer, "on the basis of the information supplied by the mathematician . . . using subroutines and its own instruction code, produces a program" on its own. Grace proposed that a smarter machine could support a less-educated programmer, and even a nonprogrammer user, as its education continued. "UNIVAC at present has a well-grounded mathematical education fully equivalent to that of a college sophomore, and it does not forget and does not make mistakes," she wrote, always the teacher. "It is hoped that its undergraduate course will be completed shortly and it will be accepted as a candidate for a graduate degree."

Compiling subroutines, rather than hand coding from scratch, removed a huge margin of human error from the exercise of programming. Subroutines were already tested, debugged, and ready to go; strung together by the compiler, programs could be written in hours instead of weeks. Grace's idea, which would be successively refined by many people in the years to follow, automated much of the drudgery associated with programming, allowing programmers to focus on the creative side of their work, and on the higher-level, systems-oriented thinking that would ultimately advance computer science as a discipline. More important, it represented something conceptually novel: programs writing themselves, a path Betty Holberton had illuminated with her brilliant Sort-Merge Generator. The idea was as appealing to marketers as it was to computer manufacturers. By 1955, Remington Rand was running print advertisements with headlines like, NOW . . . UNIVAC TELLS ITSELF WHAT TO DO!

Grace's paper ignited the computing community, but there was work to be done. Her first compiler, A-0, was elementary. "It was very stupid," she explained. "What I did was watch myself put together a program and make the computer do what I did." Although it smashed conventional programming techniques in speed 18:1, the machine code it created—the elemental, binary instruction set the computer actually executed—remained inefficient. These days, the elegance of code is mostly aesthetic, but in the early 1950s, it was still cheaper to pay programmers to manually check programs for errors than it was to burn an extra hour of computer time with clunky machine code. Within a year, Grace and her programming staff had written a slightly less cumbersome A-1 compiler; in the spring of 1953 came the A-2. With each iteration, compilers grew more sophisticated, more user-friendly; A-2 introduced what Grace called "pseudo-code," a kind of in-between language more human than computer. It wouldn't seem particularly friendly to a modern programmer, but this shorthand was the first step toward programming languages nonexperts could use. That would be Grace Hopper's legacy.

Getting the ball rolling on automatic programming was a big accomplishment, but it created an entirely new set of problems. Designing compilers and pseudo-code was beginning to feel like inventing new languages, more art than science. The new syntax had to hold the compiler together while remaining coherent to users—a language is only ever as useful as the fluency of its speaker, after all. Grace saw the proliferation of new compilers and pseudocode as a potential Tower of Babel situation, with competing, mutually incomprehensible languages crowding each other out. She proposed a few solutions. Her influential business compiler, MATH-MATIC, employed the globally understood language of mathematics as its lingua franca. Its descendant, FLOW-MATIC, assigned mathematical variables to commonly used words and phrases. These natural language commands served double duty as documentation, and their relative comprehensibility made it possible for nontechnical project managers to assess work being done. But the larger question remained: Could a single language be understood by *every* computer on Earth?

The earliest meeting to discuss this question was initiated by Mary Hawes—one of Grace Hopper's former colleagues, with whom she'd codeveloped FLOW-MATIC at Remington Rand—when she "buttonholed" a well-known computer scientist, Dr. Saul Gorn, at a conference in San Francisco, "asking if he didn't think it was time for a common business language." In April 1959, Grace called a follow-up meeting at the University of Pennsylvania, where she was adjunct faculty. Everyone in attendance agreed with Mary: the time was nigh for a shared, business-facing, hardware-independent programming language. Remington Rand wasn't the only company up to its ears in costly data installations; Grace's informal committee was full of competitors, like IBM and the Radio Corporation of America (RCA), suffering the same growing pains. But the undertaking needed a neutral—and influential—sponsor. Grace flexed some navy connections and approached the Department of Defense, which at that time was running 225 computer installations with plans for many more.

Only a month after the meeting at Penn, the DoD hosted the first organizational meeting of the Conference on Data Systems and Language, or CODASYL, at the Pentagon. Every major computer manufacturer sent diplomats to rub elbows with government brass and representatives from private industry. The common cause was a forward-thinking, easy-to-use language, preferably in simple English, that would be independent of any specific machine. Three committees were formed: short-range, intermediate-range, and long-range. The first would examine existing compilers, like Grace Hopper's FLOW-MATIC, FORTRAN, and the AIMACO language developed by the air force, to decide what worked and what didn't. After that analysis, they'd write up specifications for an interim language. The intermediate group would study language syntax and business language trends, then build on the first group's effort. Finally, the long-range group would gather all the first and second groups' research and use it to create a universal business language. A rigorously tiered process with deliverables at every step along the way: honestly, it could only have been dreamed up by computer programmers.

The committees promptly went haywire. There was no such thing as an interim language; it was far too time-consuming and expensive to switch languages at all. Grace estimated that it would cost $945,000 and 45.5 man-years—the notion of kilogirl hours being long gone from the lexicon at this point—to implement a new compiler at Remington Rand. Whatever stopgap language the short-range committee created would have to be there to stay.

"In no way was this language going to be an interim solution," concluded Grace's old Eckert-Mauchly colleague Betty Holberton, a member of the short-range committee who was, by then, supervising advanced programming at the navy's Applied Math Lab. "This language was going to be 'it' as far as I was concerned." This left the short-term group, which included three women (beyond Holberton and Mary Hawes, there was Jean E. Sammet from Sylvania Electric Products, also a Hopper protégé), to write the new business language in fewer than three months. Because programmers love acronyms, the group became known as the PDQ Committee—they'd have to move mountains of code pretty damn quick. PDQ made good on their name, delivering their findings for the new common business oriented language, COBOL, in December 1959. That following January, the specs were printed. The long-range committee was never even formed.

COBOL changed the world. The Department of Defense ran all its computer installations with COBOL and insisted that suppliers provide hardware that supported it, dictating the direction of the computing industry for a decade. Ten years after its implementation, COBOL was the most widely used programming language in the industry, and by the turn of the millennium, 80 percent of all code *on the planet* was written in COBOL, representing some seventy billion lines of code. This ultimately proved to be a huge problem: bound by the memory constraints of 1950s-era computers, CODASYL had decided on a two-digit convention to display the year, so the switch from "99" to "00" in the year 2000, it was feared, would plunge the world into chaos, in a much-ballyhooed "Y2K bug." Groups of COBOL programmers actually had to come out of retirement to squash this final and hairiest moth in the machine. It goes to show the effectiveness of Grace Hopper's efforts. The software industry, desperately needing a standard, leaped on COBOL. But although it was a language built to save the future of programming, not even its designers could anticipate just how long it would last.

COBOL was written to suit the interests of a diverse group of people; it sacrificed elegance for readability and efficiency for machine independence. It also developed a reputation for being thorny, verbose, and convoluted. Most programmers explicitly despise it. Several textbooks omit it completely. The academic computer science community, which hadn't been consulted on its design, refused to engage with it. The Dutch computer scientist Edsger W. Dijkstra famously wrote that COBOL "cripples the mind," adding that "its teaching should, therefore, be regarded as a criminal offence." Jargon File, a hacker dictionary that collected computer slang from the mid-1970s onward through various shared computer networks before first being published as *The Hacker's*

Dictionary in 1983, includes what is easily the most withering definition of COBOL:

> **COBOL: //Koh'Bol/, N.**
>
> [Common Business-Oriented Language] (Synonymous with evil.) A weak, verbose, and flabby language used by code grinders to do boring mindless things on dinosaur mainframes. Hackers believe that all COBOL programmers are suits or code grinders, and no self-respecting hacker will ever admit to having learned the language. Its very name is seldom uttered without ritual expressions of disgust or horror.

COBOL's creators wrote this off as a "snob reaction." Jean Sammet, one of the chairs of the CODASYL committee and a great admirer of Grace Hopper, pointed out that "usefulness and elegance are not necessarily the same thing." She remembered the monumental nature of the task: COBOL represents an attempt by a group of competitors to set aside their interests and create something benefiting everyone at the table, many of whom sacrificed existing projects for business or commercial software languages. COBOL was designed by committee, but it was technological armistice, a ceasefire for the good of the art.

In Jean Sammet's estimation, Grace Hopper did "as much as any other single person to sell many of these concepts from an administrative and management, as well as technical, point of view." For this, Grace is remembered as the grandmother of COBOL. Like a grandmother, she was responsible for the child but did not deliver it. Her diplomatic skills brought competitors, programmers, professional organizations, the military, and clients together. Her insistence that this milestone be reached collaboratively, rather than through competition for market share, was thirty years ahead of its time: the next generation of programmers to come of age might have sneered at COBOL's unwieldy syntax, but many would employ a similar model of distributed innovation. As her biographer points out, Grace's emphasis on collaborative development, and the network of volunteer programmers she mobilized, predated the open-source software movement by four decades. Further, building common languages that remained consistent even as hardware came and went would prove essential to the evolution of computing. If programmers had been forced to start from scratch every time a new computer came out, they'd have been playing catch-up for eternity. But automatic programming—and the foundation of efficiency, accessibility, and machine independence on which it was laid—cemented the possibility for programming to develop as a functional art form.

Beyond Grace, many women were involved in the development of automatic programming—a disproportionate amount, in fact, even relative to the demographics of their industry. Like Betty Holberton with her Sort-Merge, they developed compilers and generators. At Remington Rand, Adele Mildred Koss, a former UNIVAC programmer from the Eckert-Mauchly days, created an Editing Generator, which read the specifications of a file and automatically produced a program to transform it into different formats; the idea was refined by Nora Moser, a programmer at the Army Map Service, one of the first people to implement Grace's A-2 compiler. Moser would also help out on the PDQ committee, along with three other women, Deborah Davidson from Sylvania, Sue Knapp from Honeywell, and Gertrude Tierney from IBM. The final six-person committee to develop COBOL's specifications included two women, Gertrude Tierney and Jean Sammet, one of the first people, let alone women, to teach computer programming at a graduate level in the United States.

Admittedly, the bureaucratic labor of programming standards committees and the highly technical work of compiler design are difficult to dramatize, but for many of these women, the advancement of automatic programming represented an opportunity to assert their own importance, particularly in relation to hardware engineers. A compiler, Nora Moser noted, was like a "pseudocomputer," a machine made from code alone, and Grace Hopper observed that it constituted a "second-level" of operation beyond the workings of the machine itself, reproducible and lightweight enough to be sent in the mail. This put programmers on the same level as engineers—forging connections with symbols instead of soldering wire.

Like Grace, these women were overworked. Their employers often expected them to provide support to customers in addition to writing, testing, and debugging computer code. In an exploding industry, this was nearly impossible and, in many cases, counterproductive. But with "both the expertise to devise solutions and the incentive to make programming easier for experts and novices alike," they were in a unique position to effect real change, and did. Grace Hopper wasn't the first woman to believe in programming, automatic or otherwise. Many of Grace's female peers worked tirelessly to develop and standardize programming strategies that would transform the early computer industry, just as Ada Lovelace had made the mental leap from hardware to software a century before. But although the Difference Engine was never finished and the Analytical Engine was only imaginary, she knew, just as Grace did: a computer that does only one thing isn't really a computer.

It's just a machine.

Chapter Five

THE COMPUTER GIRLS

In 1967, the April issue of *Cosmopolitan* ran an article, "The Computer Girls," about programming. "The Computer Girls," the magazine reported, were doing "a whole new kind of work for women" in the age of "big, dazzling computers," teaching the "miracle machines what to do and how to do it." Just as a woman twenty years previous might have chosen a career in education, nursing, or secretarial work, today, its author implied, she might consider computer programming. In the photographs accompanying the article, an IBM Systems Engineer named Ann Richardson is pictured handling punch cards, flipping switches, and "feeding facts" into the computer. Looking chic in a striped, sleeveless blouse and a neat beehive, she's surrounded by faceless men in identical suits, who look down at her as she smiles brightly, a miniskirt among the mainframes.

Grace Hopper, by then in her early sixties, was back in active navy service, heading a programming languages group in the navy's Office of Information Systems Planning. Quoted in the *Cosmopolitan* article, she used one of her favorite analogies about women and programming, comparing writing programs to planning a dinner party: "You have to plan ahead and schedule everything so it's ready when you need it." It may seem like a reductive statement from someone who would eventually develop a fleet-wide tactical system for nuclear submarines, but that was Grace's style. Practical applications were the most important thing, and she always connected computers to living, breathing, everyday life. But the last word in the *Cosmopolitan* article comes from a male programmer. "'Of course we like having the girls around,' he declares, 'they're prettier than the rest of us.'"

Something happened to the generation of programmers after Grace Hopper and her peers. Although the *Cosmo* article suggests that women were being encouraged to pursue programming as an alternative to secretarial work, the field was quickly becoming far less welcoming to women than it had been even a decade before. Some estimates peg female programmers as between 30 and 50 percent of the workforce throughout the 1960s, but instead of running departments and advancing the art, they were starting to cluster "at the lower end of the occupational pool," in lower-status jobs like keypunch operator, the 1960s equivalent of data entry, punching little holes in paper cards all day long.

At the same time, technology pundits wrote often and lustily about a "software crisis" plaguing the computing industry. Due to a massive shortage in skilled programmers, software projects were coming in late, under specifications, and riddled with bugs. Many of these were dramatic, public failures: in the early 1960s, IBM delivered the OS/360 operating system a year late and four times over budget, and NASA was forced to destroy the *Mariner I* spacecraft, intended to probe the mysteries of Venus, because of a simple programming error.

Writing programs may be like planning a dinner party, but it also demands perfection on a level unlike any previous human undertaking; a single misplaced comma can send a rocket tumbling back to Earth. "If one character, one pause, of the incantation is not strictly in proper form, the magic doesn't work," wrote Frederick Brooks, who managed the disastrous OS/360 operating system at IBM. This can make programming difficult to learn. In the early decades of the field, it's also what made it resistant to the industrial production that would drive the growth of the computer hardware business: writing software is like writing poetry with the unforgiving precision of equations, and it has a practical capacity to impact human lives on an unprecedented scale.

Some historians have attributed the "software crisis" to the disproportionate development of hardware and software: as faster, brawnier computers came into use, programmers were helpless to catch up. Others have cited a personality clash between programmers—if not women, then uniquely creative, difficult, and occasionally arrogant men—and their straitlaced industrial and governmental managers. But there's a third view, one that reflects how the software crisis coincided with the long, slow decline of women in senior programming positions throughout the industry.

By the late 1960s, even as *Cosmo* was pushing programming as a neat alternative to answering office phones, women in computing were being paid significantly less than their male counterparts. In a tradition dating all the way back to those nineteenth-century human computing offices that hired women to save money, female programmers were paid about $7,763 a year compared with a median $11,193 yearly salary for men doing the same job. This wage discrimination, combined with a structural unwillingness on the behalf of computer companies to make space for childcare obligations, drove women from the industry. Meanwhile, the software crisis grew so severe that NATO called an international conference in 1968 to address the problem. No women were invited.

It was held in the Bavarian ski resort town of Garmisch. Between runs down the Zugspitze, the men attending banged out a new approach to programming, one they hoped would rein in some of the problems plaguing the computing industry. The most significant change they made, arguably, was semantic: *programming*, they decided, would heretofore be known as *software engineering*. As such, it would be treated like a branch of engineering rather than a rogue, wild-blooming field roamed by fiercely independent, self-directed misfits and women. Engineering is a job with clear credentials, not a shadowy priesthood. This change signaled a larger renegotiation of computing's professional status that would unfold through professional journals and societies, hiring practices, and certification programs throughout the 1960s and

1970s. The more the discipline professionalized, the more it grew implicitly masculine. The introduction of formal educational prerequisites for programming jobs, for example, made it harder for self-taught programmers to find employment—a change that penalized female candidates, particularly those who might have taken time off from school to raise children, above all. If computer programming "began as women's work," writes historian Nathan Ensmenger, "it had to be *made* masculine."

The shift from *programmer* to *software engineer* was an easy enough signal for female programmers to interpret. The new paradigm, subtle as it may seem, "brought with it unspoken ideas about which gender could best elevate the practice and status of programming," historian Janet Abbate writes. She argues that this symbolic exclusion, in concert with the more concrete factors at play—wage discrimination, lack of childcare, lack of adequate mentoring and support—signaled to women to avoid computing just as it was suffering from an industry-wide shortage of talent. Salt in the wound: the skills female programmers brought to the table were precisely those "software engineering" desperately needed.

The software crisis, after all, was one of deliverables. The reason projects were routinely coming in late and over budget was because they were built on rickety expectations. Getting the initial requirements right for a piece of software requires being able to listen to the client, to parse messy real-world problems into executable programs, and to anticipate the needs of nontechnical users. Despite its reputation as a discipline for introverted perfectionists, social skills are valuable in programming—even essential. Grace Hopper understood this, and it's her early self-education in a wide range of nontechnical fields that made her such a profoundly competent programmer. As she told a historian in 1968, to make the link between "the computer people" and the outside world of clients, problems, and possible applications, "you needed people with more vocabularies."

Those vocabularies aren't innately feminine, but soft skills of communication are certainly socialized as women's values. During the software crisis, aspects of software design that rely on "stereotypically feminine skills of communication and personal interaction" were "devalued and neglected," ignored by male programmers and skipped over in software engineering curricula. As a result, the industry suffered, and suffers still.

The first computers became obsolete before they became operational. The Mark I led to the Mark II, then III; no sooner was the design for the ENIAC frozen and construction begun than John Mauchly and J. Presper Eckert were inventing its successor. Those early computers had a shelf life of only a few years before something smaller, faster, and smarter came along, a pattern that has continued, breakneck, into the present day. The same can be said for programming, which leapfrogged from a tedious afterthought to an art form in less than a decade. In 1950, when the most competent adding machines on Earth required entire office floors of real estate, IBM predicted the global market for computers would be five—total. But two thousand computers were in use globally by 1960. In the 1990s, when I finally came online, IBM was selling forty thousand systems a week. Punched cards became magnetic tape as code became language, while transistors ceded to integrated circuits, and then microprocessors, miniaturizing in exponential leaps as the boxes housing them sprouted screens and keyboards, becoming household objects, and portals for work, play, and connection.

When I think of the first female computers, poring over tables of numbers in organized groups, I sense a hidden catalyst, something that seems to have ignited a sequence of events leading to our current, intractably technological condition. The women who invented programming, working as mediators between metal and mind, grew into the women who wrote the elegant abstractions of language that allowed us to talk to computers like we talk to people. Their innovations are a little harder to grasp than those that miniaturized and refined computer hardware. The ENIAC is in pieces, forty units scattered in museums across the country, but it's still a *thing,* proof of its own existence. The ENIAC programs, however, were operations conducted in time. They existed only in those brief moments when electricity pulsed through the daisy chain of patch cables strung together for the task before being unplugged and rearranged again and again.

The architects of those fleeting arrangements—the kilogirls, computer girls, operators, programmers, whatever you want to call them—changed the world. As the cultural theorist Sadie Plant writes so elegantly, "When computers were vast systems of transistors and valves which needed to be coaxed into action, it was women who turned them on. When computers became the miniaturized circuits of silicon chips, it was women who assembled them . . . when computers were virtually real machines, women wrote the software on which they ran. And when *computer* was a term applied to flesh and blood workers, the bodies which composed them were female."

With our twenty-first-century brains, we all have a shot at being as clever as Ada Lovelace, the Harvard computers, or a wartime ballistics calculator at Penn. But there's only so far we can reach before we hit the ultimate threshold—the glass ceiling over all humanity. My current machine, a top-of-the-line slice of MacBook Pro, will be obsolete by the time these words make it to ink. The machine code that Grace Hopper dreamed would someday write itself is now the engine that powers the world. It has allowed me to find the women we'll meet in this book, to e-mail them out of the blue, to wave hello to their ever-less-pixelated faces, and to make plans ending with me in their living rooms, looking at manuals, looking at photos, drinking green tea. Such is the case with world-changing technologies.

It's never easy to anticipate what they will become, or just where they will take you.

PART TWO

Connection Trip

Chapter Six

THE LONGEST CAVE

The longest cave in the world is in central Kentucky. Its limestone passages stretch four hundred miles beneath the earth in twisting patterns as intricate as the roots of the ancient hickory forests above. Within, cavers skirt bottomless pits, pass fountains of orange stone, and discover deep, icy subterranean rivers. Between the sunlit world and the depths below, white mist swirls at ankle height, like the breath of ghosts.

Kentuckians have fought bitterly to control access to the secrets of Mammoth Cave. In the early twentieth century, hardscrabble locals conned tourists into the sinkholes on their land, spurring "cave wars" that ceased only when the National Parks Department took control, evicting landowners and installing staircases, subterranean toilets, and even a grand dining room 267 feet below ground, its ceiling encrusted with snowballs of gypsum crystal. Serious cavers now enter Mammoth's wild entrances through locked grates, using keys granted by the Parks Department. They bring with them small carbide headlamps to keep warm and light the darkness.

The earliest people to map Mammoth were enslaved, installed underground by landowners to lead tours. The first of these guides, Stephen Bishop, named its features—the River Styx, the Snowball Room, Little

Bat Avenue—and discovered the eyeless white fish that swim in its deepest waters. When Bishop was sold, along with the caves, to a Louisville doctor, he was ordered to draw a map from memory. As cave maps do, his drawing looked like "a bowl of spaghetti dumped on the floor," but it detailed the nearly ten miles of passages that Bishop had discovered and remained the most thorough map of Mammoth's reaches for more than fifty years. One nameless noodle, a passage forking off the subterranean Echo River, became important a century after Bishop was buried near the cave's main entrance, his grave marked by only a cedar tree.

In Bishop's lifetime, every landowner in Central Kentucky claimed a cave entrance; if not a natural sinkhole, then a crevice blown open with dynamite. Bishop believed that all these fragments were linked into one larger system, and his instinct was shared by generations of Kentucky cavers. At the bitter end of their remotest passages, the caves *breathe:* cold air whispers even miles below the surface, and water siphons deeper and deeper into the Earth.

Proving Bishop's connection theory became the cause of the Cave Research Foundation, a ragtag group of caving enthusiasts who spent nearly twenty years linking the disparate caves neighboring Mammoth into a single Flint Ridge system. It was a family affair; once they were old enough, children who grew up playing in the woods surrounding the foundation's clapboard lodge pushed past the farthest points surveyed by their mothers and fathers. By 1972, the Cave Research Foundation had surveyed nearly every Flint Ridge lead to its endpoint, sometimes with ten-hour belly crawls through womblike tunnels. The final connection, as they called it, was imminent.

The cavers believed that Flint Ridge met Mammoth past a choke of sandstone boulders at survey point Q-87, a remote spur miles from the surface, but moving the boulders with lengths of metal pipe was backbreaking work. One expedition tried an alternate route, through a vertical crevice called "the Tight Spot." Caving humor has a nihilistic streak: the Tight Spot is a dark slit so small that only one person in the party dared enter. She was a reedy computer programmer, all of 115 pounds, named Patricia Crowther.

Pat wedged herself into the Tight Spot and came out the other end onto a mud bank. In the cool carbide light, she spotted the calling card of a previous visitor: the initials "P. H.," engraved on the wall. Back at the surface, her party kept the discovery secret. Anyone familiar with the area would know the legend of old Pete Hanson, who had explored Mammoth before the Civil War. Those had to be his initials down there, which could mean only one thing: Flint Ridge and Mammoth were connected, in a single contiguous cave spanning 340 miles. The monumental discovery would come to be known as the Everest of speleology.

Pat returned to broach the juncture ten days later. "By the way, Pat—you're leading this one," the other cavers told her. Just beyond the Tight Spot, they waded into muddy water up to their chests, until only a foot of air separated the subterranean river and the dripping cave ceiling. Soaked through and caked with mud "like chocolate frosting," they struggled to keep their headlamps dry. Eyeless crayfish skittered around their waists. When the passage opened, it revealed a wide hall, where they glimpsed the edge of a hand railing: a tourist trail in the heart of Mammoth Cave. The link was complete. Only moments before, they'd been farther afield than any cavers in history; now, weeping and falling over each other in the water, they were only a few steps away from a public restroom.

Riding to base camp in the back of a park ranger's pickup, they looked up at the stars, bright in the summer sky. Lying "in the open truck bed, with the treetops filing past overhead and falling away behind into the darkness," they contemplated their feat in silence. The long drive reinforced its magnitude: had they really traveled these seven miles underground? Their final passage, through the Tight Spot and beyond what would come to be known as Hanson's Lost River, joined the unmarked line on Stephen Bishop's hand-drawn 1839 map. After hamburgers and champagne at dawn, they slept.

"It's an incredible feeling," Patricia wrote in a journal account of the trip, "being part of the first party to enter Mammoth Cave from Flint Ridge. Something like having a baby. You have to keep reminding yourself that it's really real, this new creature you've brought into the world that wasn't here yesterday. Everything else seems new, too. After we wake up on Thursday I listen to a Gordon Lightfoot record. The music is so beautiful, it makes me cry."

The new creature Patricia felt she had brought into the world had always been there, slumbering in the darkness of geologic time. What she'd given birth to that day was not the *cave* but the *map*—not the thing, but its description. By wedging herself into the Tight Spot and bringing her lamplight to the darkness, she moved an earthly place into the symbolic Cartesian plane. Or at least that's how she might have seen it, being the party's mapmaker.

Back home in Massachusetts, Pat and her husband Will ran a "map factory," tracking the cartographic data each Cave Research Foundation expedition surfaced. Both being programmers, they brought considerable technical sophistication to mapmaking. As Pat described it, the couple typed raw survey data from "muddy little books" into a Teletype terminal in their living room, which was connected to a PDP-1 mainframe computer at Will's workplace. From this data, they generated "plotting commands on huge rolls of paper tape," using a program Will wrote—Pat contributed a subroutine to add numbers and letters to the final map—which they "carried over and plotted using a salvaged Calcomp drum plotter attached to a Honeywell 316 that was destined to become an ARPANET IMP."

The Crowthers' maps were simplified line plots, but they represent some of the earliest efforts to computerize caves, a leap in technical sophistication made possible by the hardware to which they had access: a PDP-1 mainframe and a Honeywell 316, a sixteen-bit minicomputer, both far beyond consumer-grade. Will Crowther's employer was Bolt, Beranek and Newman (BBN), a Massachusetts company that specialized in advanced research. In 1969, BBN was contracted by the U.S. government to help build the ARPANET, the military and academic packet-switching network that spawned our present-day Internet. A few years after the Crowthers used it to plot their cave maps, the Honeywell 316 minicomputer was repurposed and ruggedized to become an Interface Message Processor, or IMP—what we now call a router. These routers formed a subnetwork of smaller computers within the ARPANET, shuffling data around and translating between primary nodes, a vital component of Internet infrastructure then and now.

Will was one of the strongest programmers at BBN, and his tight, frugal code expressed his fastidious manner. A lifelong mountaineer, he taught Patricia to climb the vertical faces of New York's Shawangunk Mountains, and was known to hang from his office door frame by his fingertips while deep in thought. Will was a caver, too, and the couple spent all their vacations deep underground. "I get cold when he's not keeping me company," she wrote in one caving diary. "There's quite a draft here; the cave's breathing."

Will didn't come along on the final connection trip. He'd been at Patricia's side on earlier surveys, pushed to his limit in the underground wilderness. But the final survey fell in early September, right as their daughters, Sandy and Laura—aged eight and six, respectively—were headed back to school. One of the Crowthers had to stay home, buy the girls their books and school clothes, take them to the dentist, and register them for classes. Will knew how much the expedition meant to Patricia. She had, after all, found the lead, what cavers call "going cave," and she was dying to see it through. He told her to go ahead. He'd take care of the girls.

When Pat came home, deeply moved by the experience, Will was waiting for her. They stayed up late, holding each other and talking about the connection. When Will fell asleep, Pat crept to the Teletype terminal in the living room and entered, as quietly as she could, the bearings of the survey they'd made in Kentucky. She ran a coordinate program, and the data spooled into her hands in the form of a long paper tape. In the morning, Pat and Will brought the tape to his office, and she watched the BBN computer plot the link she'd made, beneath the earth, between two vast and lonely places. "*Now* I can sleep," she wrote.

Caving is unforgiving. Until the late 1960s, anyone entering Mammoth would have passed the glass-topped coffin of Floyd Collins, a country caver who died pinned by a boulder. Cavers become enveloped by the earth, their every move constrained by walls and ceilings of rock.

They eat very little—candy bars and canned meat—and carry their waste with them back to the surface. They have no sense of time. Emerging, they may be surprised to see the moon. As the Crowthers' friends Roger Brucker and Richard Watson wrote in *The Longest Cave,* their account of the connection trip, "The route is never in view except as you can imagine it in your mind. Nothing unrolls. There is no progress; there is only a progression of places that change as you go along."

Making the route visible is the central pursuit of serious caving. The Cave Research Foundation had a group doctrine: *no exploration without survey.* A map is the only way to see a cave in its entirety, and making maps is caving's equivalent of summiting mountains. It's also a survival mechanism. To stay safe, cavers map as they go, "working rationally and systematically to locate known passages." It's no wonder the hobby attracts computer programmers. Code is a country populated by the fastidious. Like programmers, cavers may work in groups, but they always face their challenges alone.

Not long after the connection trip, Patricia and Will's marriage deteriorated. They divorced in 1976, after a separation that left Will "pulled apart in various ways." Caving without Patricia in the company of their mutual friends in the small Cave Research Foundation community "had become awkward." Alone and surrounded by their maps, including an extensive survey of the Bedquilt section of Mammoth they'd made together in the summer of 1974, he consoled himself with long Dungeons & Dragons campaigns and late nights coding at home. When Sandy and Laura visited their father, they usually found him hard at work on a long and elegantly structured string of FORTRAN code. He told them it was a computer game, and that when he was done, it would be theirs to play.

The novelist Richard Powers once wrote that "software is the final victory of description over thing." The painstaking specificity with which software describes reality approaches, and sometimes even touches, a deeper order. This is perhaps why Will Crowther felt compelled to make one last map. This one wasn't plotted from his wife's muddy notebooks but rather from his own memories. Translated into seven hundred lines of FORTRAN, they became *Colossal Cave Adventure,* one of the first computer games, modeled faithfully on the sections of Mammoth Cave he had explored with Patricia and mapped alongside her, on a computer that would form the backbone of the Internet.

Colossal Cave Adventure—now more commonly known as *Adventure*—doesn't look like a game in the modern sense. There are no images or animations, no joysticks or controllers. Instead, blocks of text describe sections of a cave in the second person, like so:

```
You are in a splendid chamber thirty feet high.
The walls are frozen rivers of orange stone. An
awkward canyon and a good passage exit from the
east and west sides of the chamber. A cheerful
little bird is sitting here singing.
```

In order to interact with this cave, players type terse imperative commands, like `GO WEST` or `GET BIRD`, which trigger fresh onslaughts of description. *Adventure*'s puzzles are an endless shuffle of magical inventory: to pass the snake coiled in the Hall of the Mountain King, you must unleash the bird from its cage, but you can't `GET BIRD` if you're in possession of the black rod, because the bird is afraid of the rod, and in turn, the crystal bridge will not appear without a wave of the rod, and all the while you are in a maze of twisty little passages, all different—or worse, all alike.

This would have been familiar to Will's colleagues from the ongoing Dungeons & Dragons campaign they sometimes played after work. In D&D, a game with no winnable objective, a godlike "Dungeon Master" describes scenes in detail, prompting players at actionable decision points. But Will wrote the game for his young daughters. After the divorce, Sandy and Laura came to expect that they'd play computer games whenever they visited their father. According to a researcher who interviewed members of the Cave Research Foundation, "Another caver who was with the Crowthers on an expedition in the summer of 1975 reports that one glance at 'Adventure' was enough to identify it immediately as a cathartic exercise, an attempt by Will to memorialize a lost experience."

Once he'd finished coding, Will saved a compiled version of the game on a BBN computer and left for a monthlong vacation. It might have stayed there, untouched and unremembered by anyone save for the Crowther girls, had his computer not been connected to the new computer network his company had helped to build. By the time Will finished his vacation, *Adventure* had been discovered by people across the ARPANET. Where Patricia linked caves, Will linked nodes, and *Adventure,* a mental map of the long expeditions they took together, traveled wherever those links were forged.

Adventure was a phenomenon. The game was as unforgiving as caving itself. It was maddening to navigate—a "harrowing of Hell," proposed one writer who tried it—and addictive to play. Productivity in computer science labs ceased every time *Adventure* made landfall on a terminal. An *Adventure* devotee at Stanford, Don Woods, modified the code further by adding fantasy elements—an underground volcano, a battery-dispensing machine—to Crowther's austere descriptions. *Adventure*'s journey into the earth is now considered a foundational text of computer culture. Hundreds of players took their chances on it, then thousands, each scribbling their own hand-drawn maps of the subterranean world Will described.

It must have been strange for Pat. By the time she encountered *Adventure* firsthand, at a Cave Research Foundation meeting in Boston sometime in 1976 or 1977, she was Patricia Wilcox, having married the leader of the 1972 expedition. Will's computer game proved a delightful oddity for the experienced cavers in their circle, and indeed for anyone who knew Mammoth well. In Boston, the foundation spent much of their meeting playing *Adventure.* Because they played the more popular version supplemented by Don Woods, who had embellished sections of the game, Patricia didn't immediately recognize the cave it described. She told a researcher in 2002 that it was "completely different from the real cave."

Except it wasn't: Mammoth cavers who tried *Adventure* found they needed no maps. It was so accurate they could navigate it from memory. As the game spread, *Adventure* players who made pilgrimages to the *real* Mammoth Cave could scramble down the twisting passageways secure in their knowledge of the game's virtual map. A former coworker of Will Crowther's, recalling the topographical data stored on the computers at Bolt, Beranek and Newman, noted in 1985 that "Adventure's Colossal Cave, at least up to a point (or down to a point) is the same as the one in Kentucky, and the description and geology of the first few levels are consistent and accurate." That has been proven. In 2005, a group of researchers visiting the "source cave," the Bedquilt section of Mammoth, was able to document clear parallels between the cave's geology and Crowther's descriptions.

Like the fluorescein dye with which speleologists trace the course of underground streams, *Adventure*'s version of caving culture stained the entire network. Cavers seek connections, which they discover through systematic survey, collective effort, and a willingness to forge ahead into the darkness, knowing full well that when the end appears, it may be a small place, a crack in the rock so tight only the wind can broach it. The game is a set of instructions for re-creating Mammoth; those instructions explode into pencil passageways, antechambers, and pits. *Adventure* can be won only with a map, just as caves are survived only by those who know the way back out. Steven Levy, in his history of computer culture, compares *Adventure* to the craft of programming itself, writing that "the deep recesses you explored in the *Adventure* world were akin to the basic, most obscure levels of the machine that you'd be traveling in when you hacked assembly code. You could get dizzy trying to remember where you were in both activities."

I'm telling you the story of the Mammoth Cave, of Stephen Bishop and Patricia Crowther and her husband Will, heartbroken as he memorialized their adventures in code, as a way of reminding you that *every* technological object, be it a map or a computer game, is also a human artifact. Its archaeology is always its anthropology. In fact, the most famous archaeologist to study Mammoth, Patty Jo Watson, inferred an entire agricultural economy from the grains digested by the corpses preserved in the cave's constant temperature and humidity. To understand a people, we must know how they ate. To understand a program, we must know its makers—not only how they coded but for whom and why.

A half dozen turns into *Adventure,* a magic word appears on the cave wall. The real Mammoth Cave contains its share of carved messages—Patricia discovered the most important—but the "word" that shows up in *Adventure,* "XYZZY," is Will's invention. He added it for his sister, Betty Bloom, who came to stay with him after the divorce. She was one of *Adventure*'s original playtesters and a famously impatient sort. When typed, the magic word transports the player elsewhere in the game in a quick jump-flash, skipping the tedious steps along the way. According to Bloom, XYZZY, which Will pronounced "zizzy," was a family password. His daughters were told to use it if they ever got lost or needed to identify themselves. It is the original cheat code.

The first academic to seriously consider *Adventure* was a woman, Mary Ann Buckles, who compared the game with folktales, chivalric literature, and the earliest uses of film, arguing that the growing cultural importance of computer technology that *Adventure* represented would lead to a democratization of computer use "analogous to the democratization of reading that characterized the spread of printing." The literary critic Espen Aarseth, writing about the genre of forking digital literature that *Adventure* catalyzed, called it "a mythological urtext, located everywhere and nowhere." *Adventure* spawned a genre of adventure games, which mutated from text interfaces to visual ones while retaining Crowther's strange and compelling interplay of second-person description (There is a shiny brass lamp nearby) and imperative command (Get lamp). This developed into a textual physics used in virtual spaces all over the early Internet. In time, even people with no knowledge of cave adventures came to talk this talk.

Adventure has been remembered, celebrated, canonized, satirized. Crowther, who never made another game, is now considered interactive fiction's J. D. Salinger. But the domestic context from which *Adventure* emerged bears exploring, too: Will Crowther wrote the code after divorcing the woman with whom he'd mapped the cave *Adventure* emulates. It was playtested by his sister, for whom he invented the game's "magic word." It was created for the daughters he saw only on weekends and holidays and because he missed Patricia, or at the very least because she had instilled in him a love of the enveloping darkness.

Patricia Crowther had been a FORTRAN programmer at the Haystack Radio Observatory when she graduated from MIT. Like many technical women at the time, she left the computing industry behind to raise her children—and to cave, naturally. When she returned to work in the late 1970s, everything had changed. She went back to school, enrolling in all the undergraduate computer science courses the Indiana University of Pennsylvania had to offer, eventually taking a job as an instructor. In her classes, which were often attended by hundreds, she remembers seeing plenty of female students, but they would be the last generation of women to enter the field in substantial numbers. In the generation after Grace Hopper and her contemporaries, the professionalization of "software engineering" marked a sea change in the gender demographics of computing. By 1984, the number of women pursuing computer science degrees in the United States began to dive, and it has kept diving to this day, a decline unrivaled in any other professional field.

The Honeywell 316, the microcomputer at Will's workplace that would become a router on the early Internet, has one more claim to fame. Honeywell made a model for women: with a built-in pedestal and a cutting board, it sold in the Neiman Marcus 1969 Christmas catalog as the Honeywell "Kitchen Computer." It cost ten thousand dollars, came with an apron, and took two weeks of programming classes to learn how to operate, but the catalog picture shows a woman in a long floral dress unpacking a basket of groceries on top of the computer as though it were an extension of her kitchen counter. "If she can only cook as well as Honeywell can compute," the copy says, implying that the computer has "more authority, power and intelligence *than its female user*." And on her home turf to boot.

As Patricia's ex-husband's game grew in popularity, it was men who congregated around networked terminals to play it late into the night. It was men who scribbled cave maps on notepads lit by the electric glow of the screen. It was men who emerged dizzy in the light of day from each long crawl. And for all that Patricia accomplished, in the many tellings of the *Adventure* story, she has remained a background figure. Although she mapped and charted the subterranean world Will popularized with his game and made a physical leap into the unknown that few would even consider, her presence is a spectral outline of what might have been. She has been hidden in plain sight. The same could be said about many women in the early network era.

It's fitting that the networked century's inaugural collective experience would be *Adventure*. It's a story about how intimately people influence software, and how wide its impact can be. And caves were always virtual worlds, the first places where human beings experienced the ontological disembodiment we now so strongly associate with projecting ourselves on-screen. By flickering firelight or by the shudder of a CRT monitor, we see beyond the real. Symbols applied to raw granite, to canvas, to code: all of it lights up the darkness.

There's a lamp in the cave. Do you know what to do?

```
GET LAMP
```

Good. Now hold it tight, we'll need to take it with us. We'll take it through the twisting passages until they open wide to the other side and we can finally see the writing on the wall, a scrawl a hundred years old. It's our magic word, our cheat code, our jump cut through the night. You can barely read it in the carbide light: *Even when women were invisible, it never means they weren't there.*

Chapter Seven

RESOURCE ONE

It's raining in California. Levee-busting rain, water rushing to the ground like a lover kept away by drought. In the Marin Headlands, north of San Francisco, the brackish tidal plains are throbbing, egrets buzz the grasses, screeching, and I'm waiting by a flooding bus stop, raindrops on my glasses. Sherry Reson pulls up in an old Camry, opens the driver's side door just enough to poke a mop of curly hair out, and waves.

She drives me up the road to her place, a slate-gray bungalow weathered by the wind. On the way, she briefs me excitedly about the group she's gathered, which we find perched around her dining room table, eating spongy feta, broad beans, and spinach salad from wooden bowls. As I divest myself of my wet outer layers, they look up cheerily from their conversation. They've been catching up. Only recently have they reconnected with one another, but forty years ago, alongside a hundred other dreamers, hippies, and iconoclasts, they all lived together in a technological commune in San Francisco called Project One.

They'll explain to me that Project One was a mustard-yellow warehouse South of Market. Inside its eighty-four thousand square feet of interlinked habitations, they slept in hand-built bays one hundred feet wide and gathered for community meetings on the fifth floor that often ended in shouting matches and tears, or music and laughter, depending on the day. The stucco doors were tiled with mosaic patterns, there was a hot tub for communal bathing, and some residents lived in dollhouse estates of plywood and Sheetrock fastened together with nail guns. The commune's children were herded by a former marine with a foot-long beard; he paraded them around the city, introducing them to Buddhist abbots, Rinpoches, Sikh gurus, and Taoist priests.

The writer Charles Raisch called Project One a "pueblo in the city," a village peeled from the earth and turned inward until its edges met tip to tip. "It's fifty of us sunbathing and barbecuing on the roof," he wrote. "It's seven turkeys and four bands and a bowling lane size, makeshift banquet with Dumpster roses."

At Sherry's house I meet four former Project One residents who have come from all corners of the Bay Area: beyond Sherry, there's Pam Hardt-English, Mya Shone, and Chris Macie. Pam's bay at Project One was a loft bed encased in air-gapped walls of red translucent plastic. Mya slept on a wooden pallet when she came from New York City with only the clothes on her back and dreams of becoming a full-time revolutionary. Sherry inherited a little house with white steps and a front door. They cooked on hot plates, shared bathrooms, and worked in the same building they called home, which was full of political organizations, artist studios, and production facilities. But even in a warehouse full of documentary filmmakers and *Hair* dropouts, their office was stranger than most. Right in the middle, encased in a clear box of Lexan polycarbonate sheeting, they kept a mainframe computer the size of ten refrigerators.

The computer, a Scientific Data Systems 940, was one of only fifty-seven of its kind on Earth. It was easily the most valuable thing in the building, an absolute treasure to this group of hippies. Pam, Sherry, Chris, and Mya were only a few among the rotating cast of its minders, all patrons of the community computer center they called Resource One.

Like the Honeywell 316, the SDS-940 was serious hardware, and it also contributed to the backbone of the early Internet: in 1969, when it was the best computer money could buy, an SDS-940 at Stanford became one of the ARPANET's earliest hosts. By 1972, when Project One acquired theirs, the SDS-940 had become more Model T than Thunderbird, but it was still a $150,000 mainframe—far beyond the reach of twenty-somethings paying a nickel a square foot to live in a disused candy factory. And yet there it was, wired up by self-taught electricians who hauled "big noodles" of industrial power lines into their makeshift clean room. They had Pam to thank for that.

Pam Hardt-English is a deliberate, soft-spoken brunette with the studious air of a woman with well-kept secrets. She came alive at UC Berkeley, a flashpoint of the anti-war and Free Speech movements. "The school was on strike most of the time," she remembers. Even the computer science department was organizing, and when the United States bombed Cambodia, "all the computer people got together," she told a documentarian in 1972. "It was the first time many of them had ever been involved in anything. It was really exciting. We started talking about building communication networks." In the summer of 1970, she and two fellow Berkeley computer science students, Chris Macie and Chris Neustrup, dropped out of school and moved into Project One. They made it their mission to get the counterculture connected.

In a sense, it already was. The Bay Area was overrun with underground newspapers and houses with bulletin boards and free boxes in their front yards. The *Berkeley Barb* ran back-page ads for resistance organizations, and a group called the Haight-Ashbury Switchboard had even built a sophisticated phone tree in the late 1960s, linking human "switchboards" to one another to help distraught families track down their wandering hippie kids. This grew into an informal network of interest-specific Switchboards in the Bay Area, one of which, the San Francisco Switchboard, had offices at Project One. With a couple of phones and boxes of index cards, it coordinated extensive group action for quick-response incidents like the 1971 San Francisco Bay oil spill—an early version of the kind of organizing that happens so easily today on social media.

Resource One took up where these efforts left off, even inheriting the San Francisco Switchboard's corporate shell. When Pam and the

Chrises moved into the warehouse, their plan was to design a common information retrieval system for all the existing Switchboards in the city, interlinking their various resources into a database running on borrowed computer time. "Our vision was making technology accessible to people," Pam explains. "It was a very passionate time. And we thought anything was possible." But borrowing computer time to build such a database was far too limiting; if they were to imbue their politics into a computer system for the people, they'd need to build it from the ground up. They needed their own machine.

This was long before the personal computer as we know it, and long before even the microcomputer. Resource One had its sights on a mainframe system, the kind tended by experts in the large installations that had evolved from the efforts of early business programmers like Grace Hopper or Betty Holberton. Pam made a list of institutions and companies she thought might have a surplus mainframe lying around. After an exhaustive series of phone calls and meetings, she eventually cut a deal with the TransAmerica Leasing Corporation, which had a few SDS-940s gathering dust in a warehouse, one of them coming off three years of heavy use at Stanford. She convinced them by speaking the language of their common interest: the computer was worth more as a tax-deductible donation than it was obsolescing in storage. That's how, in April 1972, on the bed of a semitruck, the People's Computer came to be delivered to Project One.

That summer, while the other communards plumbed the building's twenty-foot hot tub, the Resource One group installed cabinet racks and drum storage units. Nobody on the job had done anything remotely like it—even the lead electrician learned as he went—and the software was written from scratch, encoding the counterculture's values into the computer at an operating system level. The Resource One Generalized Information Retrieval System, ROGIRS, written by a hacker, Ephrem Lipkin, was designed for the underground Switchboards, as a way to manage the offerings of an alternative economy. Once up and running, the machine would become the heart of Northern California's underground free-access network, a glimmer of the Internet's vital cultural importance years before most people would ever hear of it.

"At different points in your life, different things matter," Pam says to me. Life at Project One was joyful, fast paced, and deliriously ideological, but it wasn't comfortable. To pay her share of rent, she worked nights washing beakers in a medical lab, and when she got back, it was always cold, the concrete floors and high ceilings impossible to keep warm. On a glacial waterbed, she rarely got more than a few hours' sleep every night. The asceticism was part of the community culture, a sacrifice in service of the building's collective goals. "My brother came to live with me because he couldn't find a job," she remembers, "and he went right back to graduate school, saying, 'I can't live like this, cleaning other people's bathrooms, freezing to death at night!' But I didn't care, because I was just so engrossed in what I was doing."

That Pam managed to procure the SDS-940 from TransAmerica is still awe-inspiring. A 1972 *Rolling Stone* profile called it "one of the great hustles of modern times," citing a fellow Resource One communard who claimed that Pam, soft-spoken as she seemed, could draw blood from a turnip. In an accompanying photo taken by Annie Liebowitz, Pam leans over the open, wire-tangled back of the SDS-940 console, big glasses askew, grinning from ear to ear. No other group, in San Francisco or elsewhere, would ever manage to get their hands on such powerful hardware—let alone run it. "Pam was very driven," remembers Lee Felsenstein, a fellow dropout who followed the computer to Resource One. "She had this way of sort of screwing up her face and chewing her lower lip that certainly bespoke inner tension."

"Pamela was about the only person I have ever known," wrote Jane Speiser, a longtime Project One resident, "who was able to make a list of the fifty-three people to be contacted to get a project done, and then actually sit down and, one by one, thoroughly and painstakingly and unfalteringly contact each and every one of those individuals, by phone, mail or in person, until she got to the end of the list, even if it took three months of doing that and nothing else. She was a person of absolutely stupendous determination. There was no other way to obtain a computer (and the cost of its installation and upkeep) for a group of counterculture freaks."

Beyond securing the computer, Pam spent the better part of three years fund-raising to keep it operational. Hefty electricity bills came for the twenty-ton air conditioners that kept it from melting down. The staff needed a living, and there was always additional hardware to buy and maintain. Ironically, much of Resource One's funding came from the establishment: Bank of America supported the project, hoping to make good with—or monitor, depending on who you ask—the young people then so earnestly upending the status quo. Pam's vision to digitize San Francisco's analog counterculture wouldn't come cheap. She imagined Teletype terminals at every Switchboard phone room, in bookshops and libraries citywide, and within Project One itself, all daisy-chained into a decentralized network of shared resources and vernacular information. "If people needed something," she says, "they could type it in and get it. If they needed help, if they wanted to share a car, or needed resources, they could get it."

Basically, she imagined the Internet.

LET HER SPEAK

There's an expression in computer science: garbage in, garbage out. Fill a machine with nonsense, and it will cook it up for you without judgment, executing commands precisely as dictated. Feed it truth, and it'll do the same. It doesn't care about the signal's nature. Meaning is *our* business; the computer is a mirror that reflects us back to ourselves, and whoever controls it molds the world in their image. This might be why the counterculture's magazine of record, the *Whole Earth Catalog,* always printed the same coda on the cover of every issue: *Access to tools.*

The year Resource One installed its computer, the *Whole Earth Catalog*'s Stewart Brand pronounced that "half or more of computer science is heads." Brand was inspired by the Bay Area's constellation of forward-thinking research labs, the hacker groups gathering to play games after hours in university basements, and the scene developing at Resource One, and he wrote about computer science as the realm of mystics, sages, weirdos, and, as he put it, "magnificent men with their flying machines, scouting a leading edge of technology." With language like this, a new archetypal image of the computer user was introduced to the world. Not the studious woman programmer, like Grace, or for that matter the software engineer in suit and tie, but the wild-eyed, wild-haired hacker, who was always a man.

When Resource One finally got their donated computer up and running—its faulty drum storage unit replaced to the tune of $20K, thanks to Pam's fund-raising—they approached a general meeting of Bay Area Switchboard operators. They pitched their big plan: network the Switchboards, index all the information, and make it available to anyone with a Teletype terminal. It fell on deaf ears. The idea made no practical sense: Teletype terminals cost $150 a month to rent, and they chattered and whirred; they were *loud*. The Switchboards were human systems, organized ad hoc to suit the organizational needs of whoever staffed them. Hippie operators filed their stuff in boxes, pinned notes to the wall, each to their own. The idea of a hierarchical master system appealed to no one, practically or ideologically.

Resource One had built a library with no books. Efrem Lipkin suggested they forget about the Switchboards and install their own Teletype terminal somewhere. If people could get their hands on it, maybe they'd dig it. In August 1973, Resource One established an outpost at a student-owned record store in Berkeley called Leopold's, a hangout for artists, musicians, and revolutionaries. Efrem and Lee Felsenstein procured a cast-off Teletype Model 33 ASR teleprinter, a little worse for the wear after some heavy service at a local time-sharing company. To help the terminal blend in with its new surroundings, Lee put it in a cardboard box, padded with urethane to mute the whirring, and hand painted the words "Community Memory" on the side.

They expected that most Berkeley peaceniks, like the Switchboard operators, would eye this kind of hardware with skepticism, opting instead for the analog bulletin board on the back wall they'd been using for years. Computers were the stuff of clean rooms and lab coats, totems of a "regimented order" apprehended with "fear and loathing by members of the counterculture." The cultural work of making Community Memory approachable to the people fell largely to Jude Milhon, a notorious female hacker and writer who would later come to be known as St. Jude, patroness of the "cypherpunks," a computer subculture devoted to matters of encryption and copyright. In the 1980s and 1990s, she'd coedit the influential technology magazine *Mondo 2000.* Jude was Ephrem's girlfriend, and she'd met Lee after placing a sex ad in the *Berkeley Barb* (this was the '70s, after all). The trio got along famously.

Jude seeded the Community Memory database with provocations designed to lure users to the screen. She'd post proto-crowdsourcing questions, like: WHERE CAN I GET A DECENT BAGEL IN THE BAY AREA (BERKELEY PARTICULARLY) / IF YOU KNOW, LEAVE THE INFORMATION HERE IN THE COMPUTER. Under the table, a modem cradling a telephone handset linked the terminal across the bay to a database on the computer at Resource One. It didn't take long for the paper-and-tack message board to grow obsolete. One person answered Jude's bagel question—YOU CAN GET FRESH BAGELS AT THE HOUSE OF BAGELS ON GEARY—then another. A third offered the phone number of a local man who could teach her how to make her own. Community Memory became a free-for-all classifieds, where poets and mystics sold their wares alongside listings for carpools, roommates, and chess games.

Community Memory sprouted with unexpected uses: pseudonymous screeds, wacky come-ons, Grateful Dead quotes. It had its personalities. A guy calling himself Doc Benway, after a character in William Burroughs's *Naked Lunch,* used the service as his own alternate-reality soapbox. Doc—A DAY TRIPPER IN THE SANDS OF THIS FECUND DATABASE—developed a cult following. Jude and others began to riff alongside him. Community Memory demonstrated, long before the Web, how networked computing can strengthen local bonds and create a culture of its own. By connecting people, bagels, and jokes, it presaged the quirks of online community by a decade. "We opened the door to cyberspace and saw that it was hospitable territory," Lee proclaimed. For this, Community Memory has been amply celebrated. There's a terminal in the Computer History Museum, which also keeps Lee Felsenstein's papers, and several histories of the era cite Community Memory in glowing terms, as a "living metaphor," "a testament to the way computer technology could be used as guerrilla warfare for people *against* bureaucracies." To this, I tender two truths: it would have been impossible to execute if Pam Hardt-English hadn't brought the SDS-940, against all odds, into the hands of the counterculture.

And it's not why I've come to Mill Valley in the pouring rain.

THE SOCIAL SERVICES REFERRAL DIRECTORY

Back at Sherry's, we move into the living room, a snug beige nook anchored by low couches. I perch on the footstool of an easy chair Sherry says would suit me, being tall. The tea goes cold as everybody settles in. Sherry makes me a plate. It feels unprofessional to eat, but she insisted on hosting, and she has gathered this group on my behalf, so I pick small fingerfuls of bell pepper and gluten-free rice cracker between questions. Their memories, she has promised, will be more reliable in tandem.

At Project One, her living space often served as the unofficial hangout for the group of women Sherry considered her closest friends in the warehouse commune. She'd make food for everyone, and they'd lie around on her waterbed and talk about the building's ceaseless romantic dramas, the power struggles in community meetings, and the revolution. They all gave part of themselves to the building, into the loving chaos of the collective dream they describe to me over tea. "When I was at Project One, I was never afraid," Pam tells the group, for my benefit. "I walked alone. Nothing ever bothered me. When I left, I felt like anybody could hurt me. I had no protection. I had never experienced fear before I left Project One."

Pam left the warehouse in 1975. She felt she'd done all she could do.

The SDS-940 was by then safely installed and established. The Community Memory terminal had been relocated to a store in downtown Berkeley, and the Mission Branch of the San Francisco public library had one, too. But it might have been more than just a sense of completion that led to her move out; as the de facto woman in the original Resource One group, "Pam found herself unwittingly cast into a 'queen bee' role, with others trying to unload their emotional work onto her," Lee Felsenstein tells me in an e-mail. "This may have been the major factor in her abruptly leaving the group—the burden of the accumulated desires of so many of us to act like our mother."

The rest of the women took up her baton. Mya, Sherry, and their friend Mary Janowitz, who had been working for a different Project One organization called Ecos ("The closest thing to a governing agency in the building," Sherry explains), pooled their resources to work together on a new project, one that would nudge the Resource One computer beyond community experiments and toward social good. None thought of themselves as techies. Mary had done some punch card data entry when she was a sociology student at Barnard, finding it crushingly boring, and Mya's technological passions extended only to video art. Unlike Pam, who came from the computer science department at Berkeley, none had studied programming. But they recognized that the computer was valuable, and that it could address some unmet needs in the community. All they needed was a problem to solve.

Later, as everyone exchanged pleasantries on the threshold, Chris would turn to me. "In the beginning, it was just the men," he'd say, warmly, zipping up his Barbour. "Hippies with their old ladies, flower girls. And then two years later, the women ruled everything." This change was partially due to the efforts of the women of Resource One—to Pam's fund-raising and project management, and to the problem the rest of the women tackled after she left.

They found the perfect application for Resource One's computer through an organizer who hung around the building, Charlie Bolton. Bolton told them how social services agencies in the Bay Area didn't share a citywide database for referral information; he'd personally observed how social workers at different agencies relied on their own Rolodexes. The quality of referrals they gave varied throughout the city, and people weren't always connected to the services they needed, even if the services did exist. Chris Macie, who founded Resource One with Pam and stayed on after she left, programmed a new information retrieval system for the project, and the women started calling social workers all over San Francisco. If they kept an updated database of referral information, they asked, would the agencies be interested in subscribing? The answer was a resounding yes. The women of Resource One found their cause: using the computer to help the most disadvantaged people in the city gain access to services.

Their Social Services Referral Directory succeeded where efforts to interlink Bay Area Switchboards had failed, and for a simple reason: it actually considered its users. Social workers had no easier access to computer terminals or Teletype machines than the hippies running the Switchboards had, so the database was simply distributed on paper. For a nominal monthly fee, participating agencies received manila envelopes in the mail containing three-ring punched loose-leaf listings, organized alphabetically, to add to their own bright red Social Services Referral Directory binders. The central database itself was maintained on the SDS-940 by Mya, and later Mary, who entered the data Sherry gathered by calling local agencies. Each listing included essential information: languages spoken, service area, services provided. "We wanted the social workers to be able to do a better job," explains Sherry.

Joan Lefkowitz, a seventeen-year-old hippie kid with Janis Joplin hair and holes in the knees of her overalls, joined the group when Mya left, rounding out an all-female team. Joan had found her way to Project One after a stranger in a Santa Barbara health-food restaurant told her to knock on the warehouse window when she got to San Francisco and say "Jeremiah Skye sent you." She took the directory gig because it married her love of electronics with progressive action. Like Sherry, she spent her days on the phone: checking in on San Francisco's social workers, suicide-prevention hotlines, homeless shelters, senior centers, community groups, and Switchboards. "It really felt to me like putting the tools to exactly what they were meant to be used for, to serve the needs of people," she tells me, a cat perched on her shoulder, when I reach her over Skype. "They figured out a way to put technology to use in a way that really touched people's lives, and that just seemed completely appropriate and cool."

To keep the directory accurate, the women made dozens of phone calls a day and collated pages by manually laying them on the floor in alphabetized piles. They all dreaded the monthly chore of going to the post office with hundreds of manila envelopes, each painstakingly hand labeled. As Joan puts it, the "women didn't do the programming themselves—we just did all the rest of the work." The directory was the most useful resource that social services agencies in San Francisco had ever seen. Every month, the packet of updated listings grew, until eventually the directory became Bay Area–wide, spanning two binders, each three inches thick. Every library in the city kept a copy, as did the Department of Social Services, in all its offices. When Sherry, Mary, and Mya eventually moved out of Project One and on to the next phases of their lives, they handed the directory to the United Way. It eventually made its way to the San Francisco Public Library, which put the database in its catalog and maintained it until 2009.

Although the Social Services Referral Directory has not been included in the prevailing mythologies about San Francisco as a place where hackers and hippies came together to create the future, it mattered in more practical ways. The directory connected an unseen and pointedly nontechnological segment of the population—social workers and families in need—until well into the twenty-first century. It's not clear that the hackers, misfits, and "magnificent men" about which Stewart Brand wrote so enthusiastically would have come up with, or actively maintained, anything quite like the Social Services Referral Directory, which was an unglamorous, drudgery-intensive community service. Since its interface was a three-ring binder rather than a Teletype terminal, its true nature as a digital object remained invisible to all but those who maintained it, although the database was printed out only for the benefit of those *without* "access to tools."

Today, in the former warehouse district South of Market, buildings are still full of young people, hard at work on their vision of tomorrow. They call it SoMa now. Nobody's bumming rides—these days, the cars drive themselves. Instead of one computer, its tonnage rivaling the rooftop boiler whose installation most Project One graduates remember with equal clarity, there are computers in every pocket and on every desk. There are light bulbs South of Market smarter than the Resource One computer. The brightest minds in computer science are still here—or the most privileged, anyway—but now they're refining algorithms to sell goods and services on a network their hippie forebears could only have imagined, long haired and pointing to the box (it's an electronic bulletin board, they said; it'll help us with the revolution, they said). Perhaps if the foundational myths of the city's technology culture, which so influences the rest of the world, included things like the Social Services Referral Directory, that revolution might have had a different flavor.

At their weekly building-wide consensus meetings, the women of Resource One were used to being talked over. Sherry and Mary remember a man who wanted to keep a gun in the building, and the drag-out debates with veto-wielding holdouts. "Making decisions by one hundred percent consensus with over a hundred people—some of whom were pretty wacky—it was tiring," Joan says. They came up with a solution. Every time one of them was interrupted, the others would interject. They worked out the strategy ahead of time, no doubt during some long conversation on Sherry's bed. "We would say, 'Wait a minute, I didn't hear what so-and-so said.' Or we would say, 'Wait, let her finish.' We would do that for each other," Sherry remembers. "You're countering dominance behavior. And sometimes all it takes to do that is to wake somebody up."

The Social Services Referral Directory represents one of the earliest efforts to apply computing to social good, and it reveals what happens when the process of technological design and implementation is opened up to more diverse groups of people. When the women of Resource One—radicals, feminists, and organizers all—brought their shared values to the machine, the result was a product more beneficial to their community. They took the tools being touted as world changing by male hackers and applied them locally, making them do good in the here and now. It remains a radical idea, and it wasn't the first time that women brought such a strong concern for use to technology: pioneers like Grace Hopper and Betty Holberton, working to refine and systematize programming languages, made their craft accessible to a broader public, opening doors for even nonprogrammers to understand what computers make possible. And it would soon happen again—and again, and again—as the interconnection of computers gave women new openings to pioneer emerging fields. In the process, they'd help us all to make sense of the information age.

Chapter Eight

NETWORKS

I'm waiting for the airport bus in Marin and the rain shows no sign of stopping. I can't stop thinking about Resource One and their SDS-940. I try to imagine what it would be like to have a computer like that in my house. It had only a 64k memory—many orders of magnitude less than the phone in my pocket, which I've been known to drop unceremoniously, and have even run through the washing machine once—but I realize I'm totally jealous of Resource One. They had the computer before it got so complicated. The SDS-940 was freestanding, tethered only to a few terminals across the bay: more furniture than accessory. From San Francisco, I fly home to Los Angeles, gratefully toggling my devices into airplane mode for an hour-long respite from their constant notifications.

Computers began to be strung together at the end of the 1960s, but it wasn't until the early 1970s—as the women of Resource One were distributing their referral directory around the Bay Area and the Community Memory terminal was introducing the longhairs to the joys of electronic bulletin boards—that a skeletal version of the Internet as we know it started taking shape. At more than a dozen sites in the United States, half of which were in California, refrigerator-sized mainframes like the SDS-940 were being wired together through phone lines and routers to share resources, time, and communications. This proto-Internet, the ARPANET, was funded by the Department of Defense's Advanced Research Projects Agency. Its goals weren't social—not until e-mail came along—but utilitarian. Before the ARPANET, anyone who wanted to transfer data from one computer to another did so with a stack of punched cards or a roll of magnetic tape, and people still moved more fluidly than data. As one historian points out, before computers became accessible remotely, "a scientist who needed to use a distant computer might find it easier to get on a plane and fly to the machine's location to use it in person." The ARPANET, by linking a group of useful "distant computers" together, changed all that. With network access, a scientist at MIT could run programs on a machine in California just as easily as if they were in the room punching the keys themselves.

The ARPANET was a packet-switching network, as the Internet remains today: by breaking information into bite-size "packets" and sending it across the network in measured hops, the ARPANET insured itself against system-wide failure. If any node along the network were to go down, the packets could easily reroute themselves before reassembling upon arrival. The ARPANET's earliest users were its builders: mathematicians, computer scientists, and engineers at places like Bolt, Beranek and Newman, where Pat Crowther printed her cave maps and Will Crowther wrote router code; MIT; Carnegie Mellon; UCLA; and, up in Northern California, Berkeley, Stanford, and the Stanford Research Institute in Menlo Park. These people all contributed to designing the early Internet, suggesting new protocols, fixing bugs, and adding features as they went. Because the military and the highest echelons of academic computer science were so male dominated, it stands to reason that all these people, the first users of the Internet, were college-educated men.

Except, of course, they weren't.

A GIRL NAMED JAKE

She wore her hair combed back, parted far on the left, and chose a smart business suit for her first day on the job. She even wore heels. But when she walked into the Augmentation Research Center lab at the Stanford Research Institute, she still stuck out like a sore thumb. It was 1972, and these were Stewart Brand's "magnificent men," their hair as long and wild as their unkempt beards, sitting in beanbags, "looking kind of like unmade beds." Those who had desk chairs rolled around on the bare floor of the giant open office they called the bullpen, loud as pinballs, checking in with one another with no discernible hierarchy. Her name was Elizabeth Feinler, but everyone back home in West Virginia called her Jake.

Jake wondered what the hell she was getting herself into.

Jake was the first in her family to graduate from college. While doing graduate work at Purdue, she was so poor that she'd eat the wild squirrels her boyfriends hunted, and buy cast-off chickens from food department science experiments for five cents a pop. She found her way to Stanford through a circuitous path: after studying chemistry, she'd taken a job as an information chemist at an abstracts service in Columbus, collating scientific papers and patents into a massive repository of chemical information, one of the largest data collections in the world. She was intrigued by the sheer volume of information, and the seemingly Sisyphean task of organizing it into a useful database. Realizing she was more interested in information itself than in chemistry, she took a job at Stanford helping scholars with technical research, gathering reprints and running around to different libraries before summarizing her findings on index cards—much as a search engine works today.

At Stanford, Jake worked in a basement lab. It wasn't long before an upstairs neighbor, Douglas Engelbart, began popping down to her office for organizational advice. Engelbart had invented a computer system called NLS (oNLine System) in the late 1960s, a predecessor to the modern personal computer in both form and philosophy, and the first system to incorporate a mouse and a keyboard into its design. NLS was so visionary that the first time Engelbart presented it in public is generally known in tech history as "the Mother of all Demos." At the Augmentation Research Center, the lab above Jake's Stanford office, Engelbart's team of engineers and computer science researchers were busy imagining the future. "He would come down and say, 'What are you doing?'" Jake remembers. "And I'd say, 'What are all those people doing upstairs, staring at television sets?'"

Whatever it was, it made technical research look boring. One day, when Engelbart came down to visit her office, Jake asked him for a job. He told her he didn't have one, but he kept her in mind and, six months later, he came back. "I have a job now," he announced. It had nothing to do with index cards; instead, Douglas Engelbart introduced Jake Feinler to the wild new world of networked computing. Her life would never be the same.

In the fall of 1969, one of Engelbart's machines had been on the receiving end of the first transmission between two host computers on the ARPANET. The connection crashed halfway through, truncating that very first Internet message from `LOGIN` to the somewhat more prophetic `LO`. By 1972, when Jake joined Engelbart's team, Stanford was one of thirty-odd nodes on the growing national ARPANET; it was also home to the Network Information Center, the NIC, a central office for ARPANET affairs (everyone involved in the ARPANET in those days calls it "the Nick," like the name of an old friend with a reputation). Engelbart offered Jake the NIC, and told her the office needed to produce a Resource Handbook for the ARPANET in time for an important demonstration to the Department of Defense. "I said, 'What's a Resource Handbook?' And he says, 'Honestly, I don't know, but we need one in six weeks.'"

Jake figured it out. Because the ARPANET was initially built as a resource-sharing network for scientists, the handbook was supposed to list which machines, programs, and personnel were available to them at the various sites on the network. "It was pretty obvious what I had to do," Jake remembers. To put together the handbook, Jake made contact with every single host site on the nascent Internet—calling up technical liaisons and administrators across the country—to document exactly what they had available. Every host site was different. Colleges and universities used their host computers for all manner of things; for some, the fact that their machine was "online" was incidental. For others, it was paramount. The host computers weren't always stable. At MIT, "the kids ran the machine," Jake remembers, and undergraduates, hip to the "best game in town," often sneaked onto the network at night to mess around. Despite these challenges, the Resource Handbook that Jake's office produced became the first documentation of the Internet's technical infrastructure; weighing in at one thousand pages, it was a record of every node, institution, and person keeping the ARPANET running.

Putting the Resource Handbook together made Jake an instant authority on the ARPANET, and she eventually built the Network Information Center from a two-person operation to an eleven-million-dollar project, taking on all the major organizational responsibilities of the growing network. Working with a largely female staff, she created the ARPANET Directory, comprising, along with the Resource Handbook, the "electronic yellow and white pages" of the early Internet. In addition, she managed the registry for all new hosts, indexed the most important conversations on the network, ran the NIC's Reference Desk—a hotline for the Internet that rang day in and day out—and suggested protocols that remain core utilities of the Internet to this day.

This didn't all happen at once. After all, when Jake started at Stanford, the ARPANET was still relatively small. "We just tootled along in the background for a while," Jake explained, "until suddenly the network began expanding like crazy." As the ARPANET grew, Jake's team was responsible for keeping it organized. In 1974, the NIC took over maintaining the ARPANET's registry, the "Host Table." Every time a new institution wanted to join the network, it first needed to contact Jake's office to make sure their host name wasn't already taken and that their hardware met network guidelines. The NIC's Host Table kept the Internet running, and long before commercial outfits like GoDaddy and Network Solutions controlled the administration of Internet addresses, Jake, the well-coiffed West Virginian with a chemistry degree—out of

Jake Feinler in her office at the Network Information Center

place but having fun among the hippies and the hackers—was air traffic control, head librarian, and manager of the Internet all rolled into one.

Jake was one of only a few women involved with the ARPANET. Because it was funded by the military and built by engineers at the highest levels of academic research, the technical side of the network was dominated by men. Women arrived on the scene sideways: before Jake, a few found their way into computer networking through information science (Ellen Westheimer, who worked at Bolt, Beranek and Newman, the Massachusetts firm that designed the ARPANET's first routers, published an early version of the Host Table, and Peggy Karp at the MITRE Corporation in Bedford, Massachusetts, first suggested standardizing it), and Jake's female peers at the NIC came from all over. Reddy Dively, a NIC supervisor who managed the Host Table before Jake, was an information systems analyst from Missouri with a background in aeronautical engineering. Mary Stahl, a fellow West Virginian who'd married Jake's half brother, came on as a research associate in the early days; she had an art degree and had previously taught painting to children. One of the NIC's staff programmers, Ken Harrenstein, was deaf, and his sign-language interpreters, who were almost exclusively women, pulled double duty doing data entry for the NIC databases; none had technical backgrounds, and all learned on the fly. Their contributions to networking weren't technological so much as they were organizational. "There weren't many women who were programmers in those days, and many of the people working on networking had degrees in engineering, they'd come up through electrical engineering or physics," Jake says, "but women are very good at handling information, because they're very good with detail."

Still, they were outliers. Jake remembers being asked to make coffee in a meeting with military higher-ups, and the first time she sat down to work at one of the NLS terminals in the lab, "somebody came and yelled at me, and said, 'Secretaries aren't supposed to be using the machines.'" Such experiences shot her self-confidence. "I was sure that there was some huge plug in the sky that only I was going to trash," she worried. In due course, she worked out her strategies. Asked to make coffee, she glibly responded, "Oh I'd be glad to make it this time—maybe you'll make it the next time?" As for her fears of trashing the machine, she eventually realized that Engelbart's complex system was far more daunting to Department of Defense brass than it would have been to a secretary, or any woman accustomed to working with keyboards. "It was harder to get higher-ups to touch them," she discovered, "because they were afraid they would look foolish."

Just as previous generations of human computers working together embodied the network to come, the early Internet's female information scientists embodied another function that would eventually be taken over by the system itself: search. Long before the search engines we now take for granted existed, the NIC was the Google of its day, and Jake its human algorithm, the one person who knew exactly where everything was kept. Without the NIC's services, it was nearly impossible to navigate the ARPANET; host sites didn't advertise their resources, which were often in flux as machines and configurations changed. This left new users little to work with and put the burden of management on Jake's office. "If you didn't know where else to go," she said, "you came to the NIC." Her phones rang off the hook for nearly twenty years, and as soon as e-mail became a feature of the network, she was drowned in it. She had nightmares that she'd miss something important in the deluge. "It was just unending. E-mail was my cross to bear."

Jake picked up everything that wasn't held down—her friends joked that she never met a piece of paper she didn't like—and brought it to the NIC for safekeeping and reference. As a result, her office was full of books and papers, and her desk was a briar of loose-leaf piles so messy that one of Jake's employees once hired a cleaning person to tackle the mess. "That was the only time I ever got mad at someone that worked for me. I said, 'You can clean your office all you want but don't touch mine,'" she remembers. Jake's piles, disordered as they may have seemed, were a mental index of the goings-on of the Internet as a whole—and Jake, their chief curator and custodian.

The NIC's role in managing the ARPANET's information flow is easier to overlook than that of its builders. When information runs smoothly, after all, it can feel as natural as air. But it took a huge amount of labor to keep the NIC, and the network, operational. Jake shares a joke they told around the office, about an everyman named Joe Smith. Everybody knows Joe Smith, she explains: this person knows Joe, that other person knows Joe, too. Even the president knows him. One day Joe Smith is at the Vatican and somebody in the crowd says, "Who's that up there?" Someone else responds, *"I don't know the one in the red beanie, but the other one's Joe Smith."* Jake was the Internet's Joe Smith, she explains: totally anonymous and totally ubiquitous at once. At a certain point, she began to remove her own name from correspondence, because it just felt "ridiculous." I ask Jake when she first realized that the Internet would grow large enough to make her job impossible, and she doesn't hesitate. "That was almost from the beginning," she says. "I used to think, 'Why am I doing this?'"

The Augmentation Research Center lab went all night. The computers' capacities were as puny as the machines were giant, and it could take hours to compile a program, especially when the network was full of time-sharing users. To speed things up, researchers worked rolling twenty-four-hour shifts, running major database updates in the middle of the night, chugging coffee, Coke, and Mountain Dew to make it through to sunrise. "We were just trying to build things, get things done," says Jake, "and the machines were so crowded you couldn't do it during the day, so there was a whole cadre of people that were on the network at night trying to get work done." Sometimes they'd slump over at their desks, falling asleep directly over the terminals, but Jake and the women of the NIC retreated to the ladies' room to rest. "There was a couch in there," she explained, in an anteroom. "It was a law then—they had to have couches for women."

Much as Grace Hopper thrived under pressure in Howard Aiken's Harvard lab during World War II, Jake pulled all-nighters with the best of them. "Sometimes I worked all night long, and then came and worked all day long. It was a little hairy at five in the morning," she says. The NIC's phone lines went live at five, to catch the East Coast callers, and stayed open until midnight, and although there was only one phone when Jake arrived, by the time she left in 1989, the NIC had a bank of six, ringing off the hook with queries from around the country. It was a central hotline for the Internet: Jake's Reference Desk staff redirected callers to whomever was best qualified to answer their question, read back from lists of frequently asked questions, or else pulled information from the NIC's growing library of documentation. Still, it was never enough to keep up with demand.

Jake used the ARPANET itself to stay on top of network business. Using her networked terminal, she connected with colleagues across the lab—through screen-based chatting they called "linking"—and at sites across the country, often talking to people she'd never met in the flesh. She joined technical conversations on the Request for Comments (RFC), an ongoing interoffice memo authored collectively by researchers across the ARPANET. Although the first RFCs were print memos, once the NIC put them online, they became a shared hangout, much like a bulletin board. As the importance of the RFCs became clear, Jake, her colleague Joyce Reynolds, and a group of ARPANET researchers calling themselves the Network Working Group edited them into the Internet's official technical notes, defining conventions that we still use today. The

RFCs are relics of a time when the Internet was still small enough that nearly everyone online could be involved in a single conversation. This, of course, quickly became impossible, but it's to this sprawl—to this refractory, intractable explosion of information, connections, and people—that Jake made her most significant contributions.

None of this left much room for anything else. Jake had a hard time keeping continuity with friends outside of work, especially as the NIC's operations grew and she began to travel nonstop, coordinating ARPANET activities between Washington and its most important host sites around the country. Like many women in computing at the time, there was no question of trying to balance a personal life with her career. "I always meant to get married," she says, "but I never got around to it." Still, there was nothing quite like going out for pancakes at five in the morning after a long night in the lab, or piling into a banquette at a Chinese restaurant with a bunch of hackers to argue over who had the best pot stickers—East Coast or West. When she was inducted into the Internet Hall of Fame in 2012, she recognized how her tenacity and luck had landed her somewhere special, at a truly pivotal moment in history. "The Internet was more fun than a barrel of monkeys," she said. "Having fallen in at an early stage, I had more fun than I ever thought I would ever have." To celebrate network milestones—the first hundred hosts, the successful switch from one protocol to another—they'd throw parties in the conference room. Once, at the height of the spring season, Jake brought crabs and fresh asparagus for everyone at the NIC. "She wanted to have a crab feast," says Mary Stahl with a laugh.

"It was like my family," remembers Jake.

WHOIS JAKE FEINLER?

Jake's projects at the NIC had a knack for becoming essential. With the early ARPANET just a collection of host addresses managed by different people across the country, the NIC's Resource Handbook was, in a very real sense, the only tool for navigation. It may have been printed on paper, but it was the first Internet browser. And the ARPANET Directory, Jake's "yellow pages" of the Internet, prefigured by decades our age of searchable, reachable online social connections. Both of these simple, forward-thinking utilities would eventually become part of the ARPANET, folding the role of the NIC into the network itself. Like Grace Hopper with her automatic programming, Jake replaced herself with a machine.

As the ARPANET Directory—the Internet's yellow pages—grew alongside the network, Jake made important decisions about how it should be handled. One was philosophical: because the early Internet community was such an odd mix of military people, computer scientists, and the occasional undergrad interloper, Jake insisted that only names, never formal titles, be included in the directory. Keeping all the military rankings straight was a headache, and more important, the 1970s were turbulent times—this is the same era that the communards were claiming a People's Computer at Project One, after all. On the ARPANET, "a kid hacker would be talking to a Nobel Prize winner, and somebody that had been an anti-Vietnam 'protestor' would be talking to military guys that had just come back." Keeping affiliations out of the yellow pages leveled everyone's footing, establishing a convivial, egalitarian spirit to the Internet that stuck for decades, and allowed for the development of community over distance.

Once the paper directory grew too large to update, Jake decided to build a people finder into the network itself. She established a new server at the NIC called WHOIS. "WHOIS was probably one of our biggest servers," she explained. "We stopped putting out the directory, which was essentially the network phone book, and we put all that information under WHOIS. So you could say 'WHOIS Jake Feinler,' and it would come back and give you my name, address, e-mail address, affiliation on the net, that kind of thing." It was the original user profile. NIC staff kept the WHOIS database up-to-date with current contact information, enabling people to find one another online. WHOIS still exists: it's evolved over the last forty years, but it remains a core Internet protocol, tracking "WHOIS" responsible for any given domain, site, or service. This helps us to keep tabs on who controls what resources online, a utility that grows in significance as sources of information seem to recede ever further into deliberate obscurity. WHOIS does nothing less than keep the Internet democratic, as one policy expert notes.

The same thing happened with the Host Table. When Jake's office at the Network Information Center took responsibility for keeping and administering the ARPANET's central registry in 1974, it was just a text file: a flat, ASCII document listing the names and numerical addresses of every machine on the ARPANET, which hosts downloaded directly from the NIC. But as the network grew, the number of hosts threatened to exceed the space allotted in the Host Table, and the file itself grew too large for some smaller hosts to handle. Not all sites submitted accurate information to the NIC, either, and Jake's colleague Mary Stahl, who worked her way up from research assistant to Host Master, describes manually proofing and editing the hundreds of Host Table addresses twice a week as "a burn-out job," and an unrewarding one at that. "Nobody said, 'Oh great job, great Host Table,'" she says, and laughs. "It was always what's wrong." By the early 1980s, it was obvious that the "cumbersome and inefficient" system of maintaining a centralized Host Table was never going to improve.

The RFCs exploded with chatter about alternative systems for keeping track of all the hosts on the network. Most agreed that the new system should be hierarchical; although it seemed to be in the Internet's nature to grow in a decentralized way, it was clear to most that a sensible naming and addressing system would be essential in keeping it from descending into chaos. The community settled on a system of nation-states: they'd divide hosts into separate realms, or "domains." How hosts organized themselves within each domain was up to them, as long as they hewed to a standard addressing format, which should be familiar to anyone online today: host addresses would read `host.domain`, and users at each host would identify themselves as `user@host.domain`, the online equivalent of a mailing address with a zip code. But what would these domains be called? Jake suggested dividing them into generic categories, based on where the computers were kept: military hosts could have `.mil`, educational hosts `.edu`, government hosts `.gov`, organizations `.org`, and so forth. Commercial entities weren't part of the Internet yet, but to fill it out, Jake and her colleagues debated between `.bus`, for business, and `.com` for commercial. Jake favored `.bus`, but there were some hardware components that used the word. They settled on `.com`. That we use this domain most of all today should say something about what the network has become.

Jake wasn't a computer scientist in the academic sense, but she understood how to make sense of complex systems, and her practical contributions to the Internet all relate to building an organizational structure to give the system the best possible chance at holding together amid rapid and unstructured growth. She hired and trained a coterie of women, who worked overtime to make sure the network's core navigational tools—its manual, address book, and map—were up-to-date and accurate. In those early days, it was hugely exciting to share computing resources over a network, but with the interrelation of academic and military interests, people's natural inclination to use the Internet for social purposes, and the sheer complexity of keeping the new system online, that excitement could have very easily boiled over. Jake spent her entire career keeping the young Internet tidy, labeled, and in check; without the NIC, it very well may not have worked.

The Internet is a funny thing. Then and now, it has been a *thing:* an infrastructural backbone of immeasurable complexity, a scaffolding over modern life that has grown stronger than the building itself, which seems to have crumbled under its weight. And yet despite its inherent physicality—the routers, the interchanges, the telephone poles strung with wires, and the fiber optic cables crossing the sea—we persist in our belief that the Internet is inchoate, a cloud. The phenomenon can be traced back to its origins, to Jake's time. The hardware was built for a purpose, to share computing resources across universities and labs. But the Internet as a communications medium practically *willed* itself into being, transforming the computer from a calculator to a box full of voices. Jake, catching up on e-mails from the very beginning, could only perceive the future as it was: an information age. And information, as they say, is power.

In some quarters, the NIC's influence was seen as threatening, and was even contested. "I think there was a lot of bad feeling," Mary Stahl tells me, "about the fact that the NIC had this power. We were the source of the data." Some in the ARPANET technical community pushed back against its role as the central repository for all the network's most important documents: "They didn't want the NIC to be the be-all and end-all, because we were not the technical people."

The emergence of such power is beginning to feel familiar. What the NIC did at first was, ostensibly, administrative: the secretarial afterthought of putting the ARPANET's newly available computing resources down on paper to please its funders, and then maintaining a record of its coordinates and contacts. That information itself would take on such an outsized importance, becoming the de facto currency of the networked century, was as unanticipated as the world-changing art of programming had been a generation before. Here again were women elevating the mundane, identifying the missing human component of a complex technological undertaking. It's a little like Jake's revelation about those military men who'd never touched a keyboard before: nobody knows the system better than the operators, the librarians, and the secretaries. "The main purpose of the Internet was to push information across it," Jake says. "So there had to be somebody who was organizing the information." Who else but the women who were already there, answering the one phone number everybody knew by heart?

RADIA

There are different kinds of information or, rather, different levels of specificity over it. Jake dealt in the granular: the people, places, and things of the ARPANET. Near the end of our conversation, she tells me about the people she believed had the most influence over the young network. There were the engineers, of course, and then there were people like her, from totally different backgrounds, who handled the information side. And then there were the people who, early on, emphasized the importance of "coming up with the best suite of protocols to handle the traffic. In other words, designing the network itself." When she mentions this, I wonder if she's thinking of a woman she knows—a woman she met way back in the NIC days, on a visit to another ARPANET host site. I wonder if she's thinking of Radia Perlman.

Just when Jake Feinler was thinking of retiring, Radia Perlman was gearing up to supercharge the network's capacity to reach across the world. Like Jake, she would spend her career devising simple solutions to functionally complex problems, solutions that could scale alongside the growth of the network. Radia, who wears her long gray hair parted straight down the middle and speaks with a beatific, smiling calm, really hates it when people call her the "Mother of the Internet." Still, she can't seem to shake the title. Radia, like a lady radio.

In the early '70s, Radia was one of less than a hundred women in a class of nearly a thousand at MIT. In coed housing, she was the resident female, an oddity. She never saw any other women in her math classes. When she did glimpse one in a campus crowd, she'd just think, "Gee, that person looks out of place" before remembering how she might look out of place, too, if only she could see herself. At Radia's first programming job, the male programmers she worked with would "do things intended to be friendly and to impress me," like sitting around her as she worked, pointing out everything she was doing wrong. The dynamic made her self-conscious to the point of diverting her professional path: as an undergraduate, she designed a system for teaching programming to children using tactile controllers and buttons, inadvertently creating the field of tangible computing, but she gave it up because she worried that having "cute little kids" around would mean she'd never be taken seriously as a scientist.

Jake Feinler met Radia the first time she visited MIT, the ARPANET host site where the kids ran the computers. Although Radia was just an undergraduate, she stuck out in Jake's memory for staging a feminist action at the AI lab, where she worked alongside those nosy male programmers. "She was busy freeing the johns," Jake remembered. "The women had to go down a couple of floors to go to the john and the men's john was on the same floor as the computers. I thought that was an interesting concept, liberating the johns." Radia tells me it was nothing so revolutionary, she just put signs up, "saying something about, 'this bathroom does not discriminate based on gender, height, or any other irrelevant properties.'" She was the only woman in the lab, but they still made her take the signs down. Many years later, when she was a distinguished fellow at Intel, she'd truthfully tell visitors that she had her own private restroom. They'd be impressed, until one saw her walk into the women's restroom. "The more senior you get, the fewer women there are," she says.

Radia's mother had been a computer programmer in the age of punch card machines, and wrote one of the earliest assemblers while working for the government. When Radia was a kid, it was her mother who helped her with math and science homework, but Radia didn't inherit her love of hardware. "I wanted nothing to do with computers," Radia says. She preferred logic puzzles and music, and although she excelled in school, she secretly fantasized that a boy would beat her in math and science someday. "My plan was to fall in love with him and marry him," she says. That never happened. Instead, she was first in her class.

Radia's very first contact with computers, in high school, was through an extracurricular programming class a dedicated teacher had arranged for her. She discovered that all her fellow classmates had been taking radios apart since they were seven, and they knew fancy computer words like "input." She felt like she'd never catch up. "I never took anything apart," she says. "I would have assumed I would break it, or get electrocuted." In her introductory computer science course at MIT, the lab returned her first program with an angry note. She'd done something terribly wrong, sent the computer into a loop, and wasted reams of paper. To this day, Radia calls herself a last adopter, and although she did eventually crack programming, she gave it up in the 1970s. She has been told, again and again, the same myths we're all told about what it takes to be a good engineer: taking electronics apart from a young age and focused, borderline obsessive attention to technical details. "Certainly, people like that are very valuable," she tells me, "but they're not able to do the things that someone like me can do."

Radia designs routing algorithms: the mathematical rules determining the flow of data across a network. She broke into the field when she dropped out of graduate school to take a job at Bolt, Beranek and Newman, where she fell in love with networks but was so consistently ignored by her coworkers that she once gave an entire presentation about the solution to a difficult, unsolved routing problem, only for the man running the meeting to announce that there was a difficult, unsolved routing problem he wanted everyone to solve—the very problem to which she'd just presented the solution. She was hurt but unsurprised. Fortunately, a representative from the Digital Equipment Corporation, DEC, approached her after the meeting. "He said, 'Are you happy professionally?' And I said, 'I guess so.' And he said, 'You were just completely ignored, didn't that bother you?' And I said, 'No, I'm used to it. Everyone ignores me.'"

He offered her a job on the spot.

When she worked at DEC in the '80s, it was still so normal for programmers to chain-smoke in their cubicles that it took Radia three years to realize that she was allergic to cigarette smoke. "I always had the world's worst cold," she says. "Wherever I went, I had to carry a wastepaper basket and a large box of tissues and was just *disgustingly* blowing my nose constantly—this horrible loud sneeze you could hear all over the building. One time, I kind of walked in and I sneezed, and someone said, 'Oh, Radia's here.'" When she finally went to the doctor and realized she wasn't even sick, she threatened to quit, but they sent her back to graduate school instead. When she came back, DEC had fixed the smoking problem "by issuing a memo saying, quote, 'Don't smoke in the building,' unquote," Radia says, and laughs. Back on the job, she invented protocols that would have an indelible effect on the robustness and stability of all computer networks.

"ALGORHYME"

Radia thinks conceptually, and when she describes her approach to her work, she sounds a lot like Betty Holberton of the ENIAC Six, who liked to move around problems bit by bit, like a radar. Radia calls it "stepping outside of the complexity of a particular implementation to see things in a new way." By removing extraneous details and focusing on one thing at a time, she says, she's able to find the simplest solutions to complex problems. And because of her earliest experiences with programming, she values ease of use above everything else. "I try to design things that someone like myself would like to use," she explains, "which is that it *just works,* and you don't have to think about it at all."

In 1985, Radia was at DEC. At the time, Ethernet—a technology for networking computers locally, as within a single building—was becoming a worldwide standard, threatening to replace some of the more complex protocols that were Radia's specialty. But Ethernet could properly support only some hundred-odd computers before the packets of information traveling around the network started to collide and interrupt one another, like a bad conference call. This meant it could never really scale. Radia's manager at the time tasked her to "invent a magic box" that would fix Ethernet's limitations without taking up an additional iota of memory, no matter how large the network was. He issued this decree on a Friday, right before he was to leave on a weeklong vacation. "He thought that was going to be hard," Radia says. That very night, Radia woke up with a start and a solution. "I realized, oh wow—it's trivial," she says. "I know *exactly* how to do it, and I can prove that it works."

Like all the best innovations, it was simple. The packets couldn't all travel on the same path without stepping on one another, so there had to be unique paths between every computer in the network. These paths couldn't loop—no doubling back on the journey from point A to point B. Radia's algorithm automatically created routes for each packet based on a spanning tree, a kind of mathematical graph that connects points to one another with no redundancies. Not only did it solve the Ethernet problem but it was also infinitely scalable and self-healing: if one computer in the network goes down, as computers are wont to do, the spanning-tree protocol automatically determines a new route for the packet. This is Radia's signature touch. She designs systems that run with minimal intervention, through self-configuring and self-stabilizing behavior. This approach makes a large computer network like the Internet possible. As she said in 2014, "Without me, if you just blew on the Internet, it would fall over and die."

Radia took the rest of the weekend off after inventing the spanning-tree protocol. She wrote up the specifications on Monday and Tuesday. Her manager was still on vacation, and because this was long before people checked their e-mail every waking hour, Radia couldn't share the accomplishment with him. Unable to concentrate on anything else, she decided to write a poem. She went to the library, looking to borrow the opening line—"I think that I shall never see / A poem lovely as a tree"—from a Joyce Kilmer verse her mother had loved. "The librarian vaguely remembered the poem, too," she says, but they could find only a brief reference to it in the encyclopedia. So Radia called her mother, the former programmer who had helped her with math and science schoolwork throughout her childhood, and asked her to recite it. Her mother had an ironclad memory. "She said, 'Certainly,' and she quoted it to me over the phone." From this, Radia adapted her spanning-tree algorithm to verse:

Algorhyme

I think that I shall never see
A graph more lovely than a tree.
A tree whose crucial property
Is loop-free connectivity.
A tree that must be sure to span
So packets can reach every LAN.
First, the root must be selected.
By ID, it is elected.
Least-cost paths from root are traced.
In the tree, these paths are placed.
A mesh is made by folks like me,
Then bridges find a spanning tree.

When her manager returned to work the following Monday, he found two things sitting on his office chair: the spanning-tree protocol specifications and the poem, which has taken on mythical dimensions in network engineering circles. Many years later, Radia's son set *Algorhyme* to music, and Radia and her daughter, an opera singer and violinist, performed it at MIT's Lincoln Laboratory. As for the spanning-tree protocol, it transformed Ethernet from a limited, localized technology into something that could support much larger networks, and it is now fundamental to the way computers are networked. It's Radia's most famous, although by no means her only, contribution to networking. Her work might be invisible to the everyday user, but it's invisible in the way that laws are invisible or the rules of traffic in a busy city are invisible: it directs the flow of information at a layer beyond our conscious awareness. "If I do my job right," she explains, "you never see it."

Chapter Nine

COMMUNITIES

I like looking out the window when I fly. It's comforting to see the orderliness of the world from the sky, the way even big, chaotic cities resolve into comprehensible systems, their invisible rules made visible by a great enough vantage. Even Los Angeles, a city with no rational grid, makes sense from above: as an accumulation of town centers that grow radially, outward, until their edges touch, blending into the contiguous urban environment outsiders scorn as sprawl. As the architecture critic Reyner Banham noted, the Angeleno method of core sample—car and long boulevard—reveals a staggering, nearly geological accretion of communities, all overlaid and entangled. "I learned to drive in order to read Los Angeles in the original," he wrote in 1971.

The Internet developed much the same way, as the simultaneous expansion of many points. To read it in the original is no longer possible, although there are relics that give a taste: the RFCs where ARPANET pioneers hashed out the Internet's rules, which are archived online along with the hundreds of other documents Jake saved from the trash heap, and the entrails of Community Memory's corpus of messages, preserved as printouts on double-sided tractor-feed paper in a museum archive. But just as Los Angeles's villages enmesh in their joyful simultaneity, much of what was so wonderful about the first outposts of Internet culture has been lost in color, noise, and time.

E-mail was the ARPANET's killer app: thanks to the introduction of instantaneous messaging, the network designed for academic research became, in short order, a communications medium, with purely social chatter eclipsing resource-sharing and military applications to become the dominant use, "rather like taking a tank for a joyride," as one historian notes. This seems to be the case wherever and however computers are linked: we go online to find information, but mostly we go online to find one another.

In the late 1970s and early 1980s, the ARPANET changed hands: from the Defense Advanced Research Projects Agency (DARPA) to the Defense Communications Agency, and then eventually the National Science Foundation, growing regardless of its keeper, until it was opened to commercial interests in the early 1990s. Simultaneously, other online communications hubs began to emerge across the country, and the world, spurred by the introduction of the personal computer. Like Community Memory, these were electronic bulletin boards, but they weren't public-access terminals with a coin slot. Rather, this next generation of bulletin boards was accessed from home by early adopters with microcomputers and modems, using server software called BBS—bulletin board system.

A modem modulates and then demodulates a carrier wave before and after sending it through a phone line. It's a relatively old technology—AT&T set the standard in the early sixties—and even the modem connecting Leopold's to Resource One was a glorified telephone handset. When faster modems came along, baked into chips, the earliest BBS users filtered their waves through the finer sieve of their homebuilt computers, calling into outposts within their own area codes. The cost of long-distance calls served as a natural fence, keeping each BBS fiercely localized, at least during peak traffic hours, when phone bills added up more easily. Eager system administrators, or "sysops," improved on the original code to develop an impressive array of varietals for early PC hardware. Like its cork-board predecessor, BBS began as a system of exchange; like Community Memory, it became a world unto its own.

Some geographic communities, sensing value, invested in community networks, or FreeNets: local-access hot spots for civic information and neighborly dialogue. Many popped up in the American West, where connectivity was limited. These include the Big Sky Telegraph, a BBS network linking rural Montana schoolhouses, the Telluride InfoZone, one of the earliest Internet-based local information systems in the world, and the Boulder Community Network, which is still active today. Madeline Gonzales Allen, a systems engineer who quit her job at AT&T to build community networks after the Colorado and Utah wilderness "touched her core," developed two of these, moved by a vision of "communities coming together and deciding for themselves how they wanted to use the then nascent public Internet for the benefit of their own communities," rather than leaving it to the rarefied few.

FreeNets notwithstanding, until the popular adoption of the World Wide Web in the late 1990s, local dial-in BBS was the primary way people with personal computers got "online." As the ARPANET traveled its long road into civilian hands, BBS filled the gap. This represents nearly twenty years of network culture; during that time, some 150,000 individual BBSs flourished in the United States. Even when a meta-network, the FidoNet, came along to connect them all into a national community of villages, the local flavor of each remained distinct.

It would be facetious to claim that women were particularly well represented in this culture. In a five-part documentary about BBS, the most thorough document of its origins we're likely to ever see, a dominating percentage of subjects are male. BBS was like CB radio: a tinkerer's utopia. Some sysops made a tradition of shuttering their boards to new users in the months following Christmas, so intolerable was the annual influx of twelve-year-old boys with brand-new modems posting "Van Halen rules!"

Aliza Sherman, a developer who built some of the earliest Web sites for women (Cybergrrl.com, Webgrrls.com, and Femina.com) has a great story about coming online during this time. It was 1987 and she had just bought her first modem. She was poking around a New York City–area BBS when a sentence appeared on her screen: `Do you want to chat?` She leaped from her seat, imagining that the computer was talking directly to her. Once she'd calmed down, she realized that the message had come from a real person, a fourteen-year-old kid in Brooklyn. This was just as mind-blowing, if not more: her "first realization that there were people, not just computers, on the other side of this phone line." Those people were teenage boys.

YOYOW

Still, there were places online for grown-ups. Out in Sausalito, the same Bay Area techno-idealism that had galvanized Community Memory and Resource One a decade previous gave birth to The WELL, a BBS for West Coast intellectuals. It was a joint venture between Larry Brilliant, an epidemiologist with a computer-conferencing company, and Stewart Brand, editor of the *Whole Earth Review*. Brand was known as a connector—the counterculture had been browsing the *Whole Earth Catalog* for solar ovens, composting toilets, and radical books for nearly a generation—and a scribe of disruptive technologies. "All software does is manage symbols," he wrote in 1984.

BBS had a reputation as a realm of nerdy fiefdoms, but The WELL was different. Fans of the *Whole Earth* publications signed up to chat with the writers, editors, and subjects of their favorite magazine, expecting a level of discourse that Brand and his cohort were happy to indulge. The WELL offered membership by subscription, and the community came with tech support, a full staff ready to answer questions and troubleshoot its prohibitively complex, text-only interface, a platform called PicoSpan that takes some tenacity to learn.

The WELL was unique in the world of BBS, and its staff was, too. Stewart Brand "wanted to have an experiment," Nancy Rhine, an early WELL employee, tells me. "He wanted to see what would happen when people with community-building skills in real life transitioned into exploring it in virtual worlds." Nancy was one of those people: she came to the Bay Area from The Farm, a commune in rural Tennessee founded by a group of hippies who pitched a tent at the conclusion of a cross-country school bus caravan in 1971. Most of The WELL's founding staff was hired from The Farm: folks picking up their things and heading West after a decade plus of organic farming, group marriage, home birth, and LSD.

Naomi Pearce, a tech publicist who discovered the community in the late 1980s, says that when she first joined, "it was like somebody had literally opened a door, like I'd been in my little apartment for a while, and somebody had opened the door, and there was the rest of the world." Naomi is a Deadhead, as were many of The WELL's users in those days—Grateful Dead fans were early adopters of online communities, having traded bootleg tapes in underground networks long before the Internet. Their flavor seasoned the culture. To this day, WELL users call themselves "WELLbeings" and send "beams" for moral support. The WELL had the leggy freshness of a booming frontier or a nation determining its constitution in the afterglow of a revolution. The closest thing it had to law was an axiom handed down by Stewart Brand: *You Own Your Own Words.* Or, as the WELLbeings say, YOYOW.

When the National Science Foundation, which inherited the ARPANET from the military and rebuilt it with a faster backbone, first experimented with lifting its commercial restrictions on the Internet in the early 1990s, The WELL became one of the first commercial Internet service providers. Nancy Rhine did a little bit of everything for the growing business: she kept the books, cowrote the manual, answered phones, and handled support for international users. She'd signed up for an electronic commune, but she noticed one big difference between The WELL and her life back on The Farm, which is still famous for its school of natural midwifery. The WELL community may have been dialing in from all over the world, but it was mostly male. "The WELL stood for Whole Earth 'Lectronic Link, but it did not represent the whole Earth," Nancy says. "The bell curve of distribution would have probably been thirty-year-old white guys."

ECHO

Not everybody is a thirty-year-old white guy, and not everybody loves the Grateful Dead.

Stacy Horn, a graduate student in New York City, least of all. When she dialed The WELL for the first time in the early 1980s, she kept a wide berth from the Deadheads. There were enough conversations to interest her: with all its journalists, ex-hippies, and hobbyist computer programmers, dialing The WELL was like visiting California for the cost of a long-distance phone call. Only a keystroke away from her Manhattan apartment, this populace of bright-eyed strangers had a distinctly West Coast feel. But once she got over the thrill—and balked at her first month's phone bill—Stacy began to feel out of place. Like any New Yorker vacationing in California, the sunshine did her good, but her heart was elsewhere.

Stacy was enrolled in NYU's Interactive Telecommunications Program, a graduate program in creative technology that's part laboratory, part think tank, so whiling away a few afternoons on a BBS passed for homework. Although she was studying computers, she didn't want to talk about them when she went online. Stacy wanted to talk about literature, film, culture, and sex. She wanted a place to flirt, gossip, and argue. She wanted some women around, and friends she stood a reasonable chance of meeting in real life. Above all, she wanted something that felt like New York City—more techno-hipster than techno-hippie. "I can't send beams to someone," she complained in her 1998 book, *Cyberville*. "It's not my style. I can communicate with self-loathing, however."

In 1988, a friend on The WELL asked Stacy when she was going to start her own BBS for the East Coast. The thought hadn't occurred to her, but the relative safety of graduate school was spooling away, and she had no future plans. In the time it took for her to type a response, she'd decided: she would start The WELL for New Yorkers. She made up a name on the spot. She'd call it the East Coast Hang-Out. Echo. Stacy wasn't—her word—a "techie." Nor was she a businessperson. But she was good with people, and the tech stuff could be learned. Her time on

The WELL had taught her the basics: online communities emerge spontaneously whenever two or more individuals discover they like the same thing. People go online for information; they stay for the company. Nobody posts in a void. We share, for better or worse, together. Stacy figured that if she could get New Yorkers on a BBS and get them talking, they'd stay. They might even pay for the privilege.

She dropped an elective course and wrote a business plan. "It wasn't like I was a visionary," she says. Computer conferencing was obviously going to be huge, because everyone who tried it got hooked. "All you had to do was sit down and do it, and it was instantly fun. It was just fun, right away. You just could *see* it." But Stacy's certainty was far from universal. In 1988, beyond the Bay Area's techno-hippies and the teenage tyrants on BBS, not many people *had* sat down and done it. Web sites wouldn't exist for another three years, and there wouldn't be a decent browser for viewing them until 1993. Connections to the Internet backbone were still largely isolated to government agencies, private enterprises doing business with the government, and universities.

When Stacy took her business proposal to the bank, "people just *openly* made fun of me. And looked at me like I was the biggest loser in the world to ever think that people would want to socialize via their computers." She didn't let it faze her. Instead, she took every penny of her savings and hit the pavement, determined to build Echo from the ground up. Every night, she went out into the city, heading where interesting people congregate: parties, art openings, museums, concerts. She ducked into bars. One by one, she approached strangers to pitch them on joining her fledgling online community. Some already had computer access, but few had modems, which cost more than $100 at the time. Stacy had to convince them "to do something which sounded to them, just like it did to the bankers, *insane*."

Beyond money, there was the question of basic computer literacy: where The WELL ran on PicoSpan, Stacy built Echo in Unix, an operating system more familiar to programmers than the artists and writers she was courting. She tackled this problem with the same horse sense that inspired her street recruiting. Cool may be ineffable, but Unix can be taught. She started inviting new users over to her apartment in Greenwich Village for ad-hoc computer classes. Her students learned Unix commands and file structures within spitting distance of Echo itself, which was just a server and a stack of twenty-four-hundred-baud modems in Stacy's living room, piled high on red aluminum shelves next to loose papers and toy figurines of Godzilla, Gumby, and Ed Grimley.

Where her technical abilities ended, friends helped out. A hacker calling himself Phiber Optik debugged Echo's server pro bono; when he later went to prison for cybercrime, Echo users wearing *Phree Phiber Optik* buttons visited him once a week. Sometimes Echo would crash; sometimes things would get so bad that when the phone rang, Stacy and her handful of part-time employees would jump. *Go away,* they'd scream at the phone. *We suck.* But it was always fun. Stacy tucked toy surprises into bills, "like Cracker Jacks," until a representative from the U.S. Postal Service buzzed her intercom and begged her to stop. She was jamming the letter machines.

Echo eventually outgrew its residential digs and moved into proper office space in Tribeca, a neighborhood Stacy had often roamed back when it was a "deserted, empty, time-tripping forgotten patch of Manhattan." By 1994, Echo had two employees and thirty-five phone lines, and her user base had jumped from a few hundred interesting people she'd picked up in bars to a few thousand who'd read about Echo in the *Village Voice* and the *New York Times*. Stacy was starting to get press, and not a moment too soon—she'd nearly burned through her savings in the lean years.

She credits then vice president Al Gore for pulling Echo from the brink. "Clinton and Gore were just going around everywhere talking about the information superhighway," she remembers. Gore was lobbying for national telecommunications infrastructure—his father, a U.S. senator from Tennessee, had sponsored legislation to build the Interstate Highway System a generation previous—in the form of his High Performance Computing Act of 1991. When that legislation passed, it was instrumental in the development of many key Internet technologies, but it also had a profound cultural effect: it brought the Internet to the water cooler. Suddenly, Stacy says, "people started having this sense that there was a *thing* out there that was *important*. And if they didn't get in on the bandwagon, they were going to be left behind." The national fear of missing out made Echo an easier sell. "You know that information superhighway you've been hearing all about?" she'd say. Well, Echo is a stop along the way.

Echo merged several social functions into one relatively easy-to-use platform, what we would call today a social network. Users logged on using their real names but could post messages in threaded "conferences" on a variety of subjects using whatever pseudonyms they chose. Echo provided e-mail accounts; "Yos," real-time chats, popped up like instant messages, with three urgent beeps.

Stacy's first users were playwrights, actors, and writers. "When computer people came online and saw we were talking about opera and not games," Stacy explained in 2001, "they left." Marisa Bowe, a longtime Echo user, remembers people on Echo as funny, snarky, and smart. "There was a contingent of people who would never be able to handle themselves at a party," she says, "but it was online that you could see how brilliant they were." They were artists, liberals, programmers—the kind "who could actually talk to people"—some media types, and a heavy nonreligious Jewish contingent: the New York intelligentsia. When Echo reached peak trendiness, it saw its share of celebrities, too, like magazine writers Rob Tannenbaum and James Walcott, the screenwriter William Monahan, and even John F. Kennedy Jr., who posted as "flash." Stacy helped him set up his account in her apartment, as a thousand incredulous Yos pinged her screen.

Affectionately, Stacy called her users Echoids, after a catchphrase from the Yogi Bear cartoon she remembered from childhood: "Heavens to Murgatroid!" The terse, campy name was suitable for a group of East Coast wiseasses who thought of their online community as a virtual salon but weren't above gabbing about TV. While The WELL's founders, Stewart Brand and Larry Brilliant, were both accomplished figures in the West Coast's tech and cultural scenes, Stacy Horn was a "punk rock suburbanite-city girl who didn't do a heck of a lot" until she started

Echo. The difference between the two communities started there. "West Coast/East Coast, boy/girl, night and fucking day," she wrote in 1998.

Journalists who covered Echo in the 1990s tended to focus on its uniquely New York flavor. "Sometimes newcomers don't realize that if Echoids attack their views and mercilessly beat down their arguments without so much as saying hello, they're not being hostile," a 1993 *Wired* profile explained. "Far from it. It's just that special New York way of saying, 'Welcome to our world!'" The quintessential Echo thread was *I Hate Myself.* That's where Echoids popped in, sometimes daily, to add to a growing list of their own trespasses: "I hate myself for being a fucking addict," one user posted in February 1992. "I hate myself for letting my Chia Pet die," quipped another.

AND NONE OF THEM WILL GIVE YOU THE TIME OF DAY

Stacy Horn had a catchphrase for Echo. She used it in print ads, and once the Web came along, she put it on the front page of the Echo Web site: "Echo has the highest percentage of women in cyberspace—and none of them will give you the time of day."

Today, women dominate social networking platforms like Pinterest, Facebook, and Instagram, but almost no online services in the 1980s had a significant female population or made any effort to cultivate one. At the time Stacy founded Echo, the entire Internet was only about 10 to 15 percent female. But women made up nearly *half* of Echo's user base. "My success was due in part to the fact that I was the only one trying," she explains. Just as she had when courting her first users, she went up to women everywhere she went and conducted informal interviews about their online experiences. If they hadn't had any, she asked what kept them from it, and when they answered, she listened. She made diplomatic entreaties to local women's groups and gave the editors of *Ms.* magazine their own Echo conference, which exploded with frank conversations about menstruation and body hair that Echo's men read in awed silence before creeping in with questions they'd probably been waiting years to ask. She started a mentoring program, and even made membership on Echo completely free for women in 1990.

Her advocacy was personal. While still in grad school at NYU, she'd worked as a telecommunications analyst for Mobil, the oil company. At work, she was responsible for making sure regional Mobil offices were linked to data centers in Princeton and Dallas. Once she woke to the possibilities of online networks, she brought her passion to Mobil. She was convinced that employees working on national data network installations would make fewer mistakes—that fewer things would fall through the cracks—if everyone could discuss what they were doing, as they were doing it, with one another, in real time. "We would have these meetings, where it would just be this long conference table with everybody in corporate telecommunications," she says, "and it was just me and a bunch of men. I would get up, and I would try to promote the idea of social networking . . . and every time I would do this, they would basically just try to shut me up. I just kept trying and trying and trying, and they would just shut me down."

At the time, Hillary Clinton was lobbying to push health-care reform through Congress. Stacy watched her on the news, ducking and bobbing as rooms full of men tried to undermine her project. She looked so resilient. "She was a master at it, even then, just deflecting and not getting angry," Stacy remembers. The image affected her deeply: if Clinton could deal with male indifference on a national scale, Stacy could prove her bosses at Mobil wrong. When Mobil moved her group's offices to Virginia, she decided to stay in New York. Her severance payment covered the first Echo servers, and the experience became a bellwether for its culture. Stacy had been the only woman in the boardroom, and she was going to make sure she wasn't the only woman in the chatroom.

She didn't always win. When Aliza Sherman—the developer who thought her computer was talking to her the first time she went online—signed up for Echo, she couldn't wrap her head around the culture of the place. Like Stacy surrounded by Deadheads on The WELL, it just wasn't a fit, so she canceled her account. Stacy called her personally. "This is Stacy Horn," Aliza remembers her saying, "I saw that you were leaving, we need more women here, don't leave, what can I do to make it easier for you?" Later, Stacy sent her a letter. Aliza left Echo anyway, but she kept the letter. It's a physical reminder of a time when every woman online made a difference.

Stacy's ex-husband used to describe her as five feet of Buddha nature, an immovable force for good in a changing world. "In those days," Stacy writes, "journalists wrote that I started Echo to provide a safe place for women on the Net. Bite me. I wanted to get more women on Echo to make it better."

EMBRACEABLE EWE

Echo's biggest conferences—Culture, Media, Death—were like Greek agorae in cyberspace: open public assemblies where anyone could be heard, provided they spoke loudly enough. Although these conferences were often the first stop for new users and wizened Echoids alike, Stacy knew that even the most democratic public spaces have limitations. "I talk differently when I'm with my choir friends, I talk differently when I'm with my drummer friends," she explains, "and if I'm in a group of all women there are things that I'm going to say that I won't say in a mixed-gender group." If Echo was going to be an extension of the real world, fulfilling a variety of emotional needs, it had to have both common areas and private spaces.

This is one of the most interesting, and most prescient, aspects of its design. Sometimes we love to grandstand to an audience of strangers at a party, but we might share more personal opinions over dinner with five or six friends. Those two ways of communicating aren't mutually exclusive; in fact, having both makes us whole. In the landscape of present-day social media, there's just as much action in private Facebook groups, Slack channels, group text threads, and direct messages as there is on our public feeds. We need private space online; Stacy recognized this truth early. If someone in a public conference was being annoying, for example, regulars were more likely to want to take the party back to their place than they were to log out entirely.

Stacy built the possibility for private conferencing into Echo from the outset, and Echoids ran with the idea: there was a private AA group for recovering addicts, and one for users under thirty. The sex conference was twenty-one and over. The Biosphere, named after the hermetically sealed biological experiment in Arizona, was private for the sake of being private. Another, Women in Telecommunications (WIT) was the female-only corner of Echo. This one was Stacy's baby, and she policed it strictly, granting Echoids access to WIT only once she'd spoken to them on the phone to ensure, as best she could, that they were women. An imperfect process, but regardless a layer of real-world scrutiny hard to imagine today.

WIT was Echo's powder room, the place where female Echoids sneaked off and talked among themselves, vacillating between sex advice and politics, dating and personal trauma. In a thread called "Is

Someone Bothering You on Echo?" they'd report instances of abuse and harassment and compare notes about online creeps. Not everyone liked WIT; some female Echoids found it cheesy, antithetical to the dark humor that had drawn them to Echo in the first place. With Stacy's blessing, a couple of them started BITCH, an invitation-only hangout for girls with attitude. Some called it "WIT with a leather jacket." Marisa Bowe, who "detested the syrupy nicey-nicey BS of WIT" compared BITCH to a "sleazy dark dive in the very lower very east side." If Echo was a digital overlay of New York, then every corner of the city had to have its analog somewhere, and sometimes a girl just needs to get a drink, talk trash, and blow off some steam. Stacy set up a similar conference, MOE—Men on Echo—for those guys excluded from conversations on WIT and BITCH. This quieted cries of preferential treatment for women, but it left transgender Echoids in the lurch.

The problem didn't become apparent until 1993. That's the year Embraceable Ewe, a trans woman, requested access to WIT. Nobody quite knew what to make of it. The question was new in cyberspace, and female Echoids took to WIT to share their opinions. Some said let her in. One agreed, on the condition that she avoid the conversations about PMS; another questioned why the same condition wouldn't apply to a postmenopausal woman. Still others argued that WIT was a space for women who were *brought up female,* and that a trans woman, having benefited from the systemic advantages of the patriarchy for at least some of her life, wouldn't share that history. The conversation became a sprawling consideration of gender in cyberspace, the first of many to come. Somebody suggested starting another private conference for trans women. "Shades of separate but equal," worried Stacy.

Echo was a stronghold for New York lefties and artists. It had an LGBT conference, Lambda. It's jarring to read about how fiercely Echoids debated the Embraceable Ewe question, because it reveals just how much understanding of the trans experience has changed in the last two decades. But another factor at play that might be invisible to contemporary readers is an ongoing context of female impersonation. The text-based Internet's social spaces were rife with gender crossing, and with the shortage of women online, men posed as women with regularity. Aliza Sherman discusses this phenomenon in a first-person account recorded for the Women's Internet History Project. "Back then it was just so odd: so many men pretended to be women online. They would have these websites and you would think they were female, but they were actually men posing as women. I'm not going to go into the possible reasons for that, but it was really hard to find actual women with websites."

The phenomenon stretches back to some of the earliest social spaces on the Internet, text-based fantasy games called Multi-User Dungeons, or Multi-User Domains. On MUDs, gender play was encouraged; most MUDs provided a long list of gender options for new characters. Pavel Curtis, the designer of one of the most well-known social MUDs, LambdaMOO, observed that its most sexually aggressive female-presenting characters were often played by men "interested in seeing how the other half lives, and what it feels like to be perceived as female in a community." The media theorist Allucquére Rosanne Stone called this "computer crossdressing." In a 1991 paper, she cites another example: Julie, a beloved message board personality who turned out to be a middle-aged man. The false Julie had been mistaken for a woman the very first time he went online, and he was so fascinated by the way that women speak to one another in the perceived absence of men that he kept up the charade for several years, building an entire fictional persona for his feminine alter ego. This was a regular occurrence on BBSs, Listservs, Multi-User Domains, and other chat platforms throughout the 1990s.

"On the nets," Stone writes, where "grounding a persona in a physical body is meaningless, men routinely use female personae whenever they choose." This cut both ways, of course. In the relative obscurity of the textual Internet, anyone could try on a new identity, which had its creative advantages, and allowed a great many people the firsthand experience of previously invisible gender dynamics in a group setting. Women could choose male aliases in order to avoid undue attention or harassment, and trans people were able to express their gender identities safely and freely. However, one practical effect of all this computer crossdressing was that the few women online had a much harder time finding one another.

For many Echoids, the subtext of the Embraceable Ewe debate was, unbelievably, this: that if a trans woman joined WIT, she'd be followed by men "pretending" to be women. It wouldn't matter who Embraceable Ewe was or whether she deserved a safe space herself, because the floodgates would be open. "I didn't know what to do," Stacy tells me. "My fear was that if I let her in, all these men would start saying 'I'm a woman, let me in,' and how would I know who was a woman who was not?" Stacy wanted to allow for as many forms of communication on Echo as possible. Within WIT, that meant a place without male voyeurs.

"I felt like the George Wallace of cyberspace," she wrote. Unsure how to resolve the problem, Stacy told Embraceable Ewe she could have access to WIT once she'd had her gender reassignment surgery. She retracted the mandate a few months later—and to this day regrets ever making it—but Embraceable Ewe left Echo anyway. In the following years, several other trans women joined. The community grew more familiar with their identities. Echoids who had been clueless about the trans experience began to understand. "Cyberspace makes it easier to *hear people out,*" Stacy wrote. Trans Echoids shared their stories, pushing back: Why should an expensive surgery be required to prove who you are? What authorizes the medical establishment to determine gender? Echo adapted its policies. "Echoids were only able to reach a tentative understanding and agreement . . . through a lot of words, volumes and volumes, over years of time," Stacy wrote.

PERRY STREET

When I first make contact with Stacy, she's busy transferring Echo, which still exists, from one server to another. Moving a thirty-year-old online community is like transporting a sourdough starter: Echo is a living culture in a jar. Susceptible to the harshness of the outside world, it must be fed and kept warm in order to provide a link between the old and new.

The big transfer falls on the same weekend Stacy and I are supposed to meet in New York. Late the night before our coffee date, I get an e-mail: the server move was a disaster, and she's been working all day and night to fix it. The crisis might spill into tomorrow. An hour later she writes me again. "All right," she begins. "This move is going from bad to worse." We won't be able to meet, after all. Apologetic, she sends me a few addresses: the original apartment, in Greenwich Village, where Echo's servers first lived, and the Art Bar, just around the corner, a frequent Echoid haunt.

I visit both places on a crisp fall morning. I'm not sure what I hope to see, or feel, when I lay eyes on them. Although bound by a physical backbone most of us hardly ever consider, the Internet seems to float, boundless, from device to device. The sum of its experiences can't be revealed by any consideration of its infrastructure, and a cable tells us nothing about the conversations coursing through it. I stand on Perry Street, squinting as I hold the map on my phone up against the front door of the apartment building that once housed the city's first social network, and to be honest, I don't feel much.

It's a nice building. Like many in this part of Manhattan, it looks expensive, with windows framed by decorative moldings of lions and mythological griffins. The street is lined with spindly, leafless trees and iron streetlamps; bustling around me are athletic men with dogs, tourists cut loose from the pack, women walking with frozen yogurt, all oblivious to the significance of this place. Shouldn't there be a plaque? We do that for buildings where poets and painters lived or died. Why not software and servers? I am suddenly moved by the idea: here lived Echo, where New Yorkers loved and laughed and complained and told one another just how much they hated themselves, and why.

When we finally do get a chance to talk, Stacy tells me a story about Perry Street. She has an expressive Long Island accent cemented by three decades as a rent-controlled Manhattanite. When she speaks, you can hear the italics. Telling me about how her neighborhood has changed over the decades, she inflects at pointed intervals. "There were *laundromats, delis,* and all of that is gone," she says. "I have to walk *forever* to get to the laundromat, which is a *drag.*"

In 1990, back when Greenwich Village was still a *neighborhood,* Echo was still a server in Stacy's apartment. Thanks to Al Gore's information superhighway, however, business was booming, and since Echo's modems relied on local telephone service, Stacy's incoming requests were jamming all the phone lines in the neighborhood. The phone company was forced to provide Stacy with a dedicated line, but in Manhattan, cables run underground—the city is too dense and tumultuous for overhead telephone lines—and so New York Telephone dug up Perry Street, all the way from the local central office to Stacy's doorstep. "It was this weird thing," she tells me, "to see the streets of New York being *ripped up* for something that I was doing. Not only was it a proud moment for me, it was like seeing the future happening, and knowing that I was helping."

Stacy's neighbors didn't share her pride. Their service was being interrupted, their street torn to pieces. Stacy knew they'd thank her later, when Perry Street had the fastest Internet in New York City.

f2f

According to a systemwide poll conducted around 1998, 83 percent of Echoids said they met regularly face-to-face. Echo hosted monthly meetups at the Art Bar and a monthly softball game in Central Park. Stacy started a reading series for Echoids called Read Only; she organized public talks, a film screening series, and the Virtual Culture Salon, a bimonthly event copresented by the Whitney Museum. By the mid-1990s, New York's "Silicon Alley" would be overrun with mixers, meetups, media events, and dot-com hype, but Stacy was the first person to link New York's technology pioneers together. Echo was, as an editor of the online newsletter *@NY* wrote in 1996, "the bedrock of Silicon Alley."

Stacy never really saw it that way. Echoids didn't network—they jammed, explored, went sailing. A group of Echoids got together monthly to divvy up a side of beef arranged for by a cohort upstate. Echo's house band, White Courtesy Telephone, played regularly at Lower East Side clubs. When Echoids had poetry readings or open-mike appearances, they listed their gigs so others could show up. And after they got home, the first thing they'd do was log back on to Echo to swap stories. "The strongest virtual communities are not strictly virtual," Stacy explains.

If nothing else, real-world accountability made Echo civil: it's harder to be cruel to someone online when you might run in to them on Monday night at the Art Bar. Echoids saw the whites of each other's eyes regularly enough to know one another as people rather than usernames. Reconciliations were made over a beer or a friendly game of softball. For better or for worse, Echo was a community, and it took care of its own.

When that didn't work, Stacy stepped in. A moderator in every sense of the word, she went above and beyond to make sure people were getting along. If Echo was a small town, then Stacy was the mayor, sheriff, and the tourism bureau at once. She made the rules and enforced them. If somebody was being awful, she had the authority to give them the boot. "Stacy was more of an autocrat than these huge corporations can be,"

"A woman started the business, half the hosts were women, so just my doing that alone gave everything a different feel the minute you logged in. They felt more comfortable."

Howard Mittelmark, a novelist who has been active on Echo since 1989, tells me. "She could finally say, 'You—you're fucking out of here.'"

Banishment was reserved for only the worst offenders. There's an entire chapter in Stacy's book devoted to what Echoids called "the Fear": that feeling of horror a truly objectionable person sows with their behavior online. Echo had a Nazi once, and its share of sexual harassers. Marisa Bowe remembers an episode in which a young guy—"who thought he was, like, really subversive"—posted about incest in every Echo conference. Stacy forbade ad hominem attacks, and since the incest posts weren't directed at anyone in particular, they didn't violate Echo rules. Marisa found them as upsetting as vandalism. "When the world that you're in is made *purely* of speech," posting offensive material just for the sake of it is "like you're bombing the buildings."

All of this should be familiar today. The Internet's still got its share of Nazis, trolls, and stalkers. And a social network that folds seamlessly into everyday life feels like a natural idea. That's what all our social networks *do* now: we invite our friends to parties on Facebook, and we show them what they're missing through photo and video stories. We follow one another's travel schedules, love lives, and pets on social media, moving with ease from real life to online life, blurring the boundaries between the two and often mistaking one for the other.

But Stacy understood early on just how important people are to the network; as Echo's final authority, she nurtured discussion and enlisted Echoids to lead conversations in their fields of interest. These "hosts" had carte blanche to engineer the particular cultural atmospheres of their conferences. "Echo is Echo because of the hosts," she wrote in 1998. "The relationships we have are formed by what we tell. Hosts get us to tell each other everything." In order for Echo to thrive, Horn realized that it needed a core base of vocal, participatory users. Howard Rheingold, in *The Virtual Community: Homesteading on the Electronic Frontier*—a book about virtual communities that completely omits Echo, incidentally—documents this strategy at work across early online communities, from a BBS in France with paid "*animateurs*" culled from its most active users to The WELL hosts in his own backyard. "Hosts are the people," he wrote, who "welcome newcomers, introduce people to one another, clean up after the guests, provoke discussion, and break up fights if necessary."

We're all hosts now, the comment fields below our Instagram and Facebook posts our own personal conferences. But the more formal convention continues as well: Reddit, the so-called "front page of the Internet," is a glorified bulletin board system, and each subreddit—which Echoids would call a conference—has its moderators, who make editorial decisions and lead conversation. Communities like Facebook and Twitter have moderators, too, although the role is no longer practiced by deputized users. Instead, paid contractors, working in obscurity, often abroad, manually remove offensive content and respond to claims of abuse and harassment. That these after-party cleanup duties would become more important than the party itself is an inevitability of scale. It's what happens when you invite everyone on Earth.

As a rule, every conference on Echo had a male host and a female host; the pair, in exchange for free access to the service, would keep things civil and interesting. That half of Echo's hosts were women was, again, a conscious effort. Stacy called it "cyberaffirmative action." It was essential to her that when women dialed into Echo they'd see other women in the thick of it. Knowing they were not in the minority, and that their administrators were not all men, encouraged women to post instead of lurk, and in that way they became part of Echo's culture. The men had to hear them. "I heard women talking about things that I wouldn't normally hear women talking about," says Howard Mittelmark, "talking about men in ways that they wouldn't usually share with me."

At its peak, Echo had four to six thousand users. More than half were lurkers. Even today, most people are lurkers: on social media platforms everywhere a vast silent majority listens, reads, and keeps to itself. It's the social dark matter of the Internet, the force holding us all together. Those who *did* post actively on Echo—about a tenth of Stacy's subscribers—were the diehards. They were the hosts, the contrarians, the pot stirrers, the core group. Although Echo still exists, tumultuously but successfully moved over to a Linux server not long ago, only a fraction of that core group remains.

Like any family, the remaining Echoids have gone through a lot together. There has been at least one Echoid death. Echoids were among the earliest to discover how difficult it is to mourn someone when their words remain, like ghosts, in the machine. Two decades of conversation represents a staggering amount of emotional equity: time spent arguing, riffing, making nice, making opinions, and flirting. Echo has had its share of romances, and many Echoids went all the way to cybersex in the service's private corners. And then, of course, there is the internecine drama of such affairs gone south.

But Echoids share a larger cultural history, too. During O. J. Simpson's infamous 1994 Bronco chase, while most of America watched the helicopter footage on the news, Echoids were posting their immediate reactions in real time. Stacy shares part of this thread in her book, *Cyberville,* and it reads exactly like a group of people live-Tweeting a major cultural event a decade before the invention of Twitter. Echoids called it "simulcasting," and it came naturally to the medium. They did it during the Oscars, during Anita Hill's testimony at Clarence Thomas's confirmation hearings, during the 1993 World Trade Center bombing. On September 11, 2001, Stacy ran to her keyboard at precisely 8:47 A.M. and typed: A PLANE JUST CRASHED INTO THE WORLD TRADE CENTER.

On Echo, discussions ranged high and low. People shared the contents of their fridges, their purses, and their pockets. "The hottest topic for a while was a discussion about shampoo," Stacy wrote. "Shampoo! Who would want to talk about shampoo? And yet the conversation is oddly revealing." Details like these humanized the words on the screen: here were New Yorkers living their lives, eating and complaining and washing their hair, while the world carried on around them, as it does around us all.

Stacy recently donated all of Echo's archives, its twenty years of unbroken conversation, to the New-York Historical Society. "Someone in the twenty-second century and beyond is going to look back and have this treasure trove of history," she tells me, genuinely proud. Echo's legacy might not be what it predicted about the future of social media—although it did predict basically everything—but what it will reveal about its past. It provides an account of how real New Yorkers lived, unvarnished by the passage of time. There's a reason Stacy's book about Echo is called *Cyberville:* Echo was an elective community reached through keyboard and modem. It had its hotshots and hierarchies, its cops and its creeps. Not everybody got along, but most stayed for the camaraderie. They were citizens of a city within a city where you were only as good, or as alive, as your words.

Stacy Horn originally sold Echo to New Yorkers as a local stop on the information superhighway, but the highway was built through town. Echo became like one of those rest stops on the scenic route, full of characters and self-published histories. People wander in every once in a while, but mostly it's off the map, somewhere in the collapsed space between the Internet's past, a Wild West of homesteads, and today's centralized, corporatized Internet, which is many things but certainly not a conversation where anyone writes for free, or freely. Echoids get on the highway sometimes to visit big cities like Facebook and Twitter. For all the action in these glitzy new metropolises, it all feels eerily familiar.

To join Echo today, you fill out a form on the Echo Web site and wait for Stacy to send you a new-user packet in the mail. Mine took a week. It contains a welcome letter, return-addressed Echo Communications Group, which begins with an exhortation—"Thank you and congratulations! You're about to join an eclectic, opinionated, slightly dysfunctional community called Echo!"—before moving on to the caveat: "We use this totally retro software that we haven't upgraded in forever and getting around on Echo is not like getting around on the web. But hang in there. It's worth it." The enclosed instructions might be millennial-proof.

Echo is still a BBS, which means it's not on the Web. Rather, it's accessed via telnet, a protocol that allows me to dial Stacy's server directly. This requires opening Terminal, a command-line application that serves as a kind of text-based window into the operating system, and summoning Echo with a typed command:

```
$ ssh claire@echonyc.com
```

The Echo welcome packet also includes a cheat sheet of Unix commands, and mastering those takes an afternoon. I am ostensibly savvy about these things, but using Echo feels like how hacking looks in movies. Eventually I get the hang of it: typing `j mov` to join the Movies & TV conference and `sh 222` to read the two-hundred-and-twenty-second item posted there, I read Echo's ongoing *Star Trek* discussion, which spans a decade. When I type `O`—as in "who is online?"—I discover the handful of people here with me in Stacy's server, where there are no advertisements and no clickbait. Using Echo feels like slicing a knife through the desktop metaphors, through the illusions, through the cloud. It's a pure channel, the skeletal core of social media.

For those of us raised on the semantics of the World Wide Web, systems like Echo are deeply unintuitive, because the Web is our only frame of reference for how the world's digital information is traveled. Even after all I've learned about Echo's world of text, while I'm connected to it, my fingers keep drifting to the trackpad, so hardwired am I to point-and-click interfaces. Echo is not on the Web—Stacy couldn't afford to make the leap back in 1993—and its "totally retro software" feels shockingly unfamiliar. With no links, nothing to click, and no URLs, it's missing something fundamental to the modern experience of communications media, something so intimately familiar that many people don't even notice it's there. If we're all fish in the online sea, Echo is missing the water. It's got no hypertext.

Chapter Ten

HYPERTEXT

When we say the words "Internet" and "Web" today, we often mean the same thing: the force, larger than nature, that emanates from our screens. But "Internet" and "Web" are not interchangeable; as we've learned from Echo, people were connecting online for decades before graphical Web pages appeared on the scene. They dialed directly into each other's machines and into host computers to exchange files and post messages, as on Echo or The WELL, or they participated in commercial online services like AOL, CompuServe, and Prodigy. Before the Web, when people talked about the "Net," or "going online," this is usually what they meant. Many of these ad hoc networks interacted with, and eventually coalesced with, the infrastructure of the Internet, which finally hit critical mass when the ARPANET's successor, the National Science Foundation's NSFNET, gave way to the network we use today, with the appearance of commercial Internet service providers, in 1994.

It was a great complication. Early maps of ARPANET were easy to read: a few nodes placed in America's academic and military capitals radiating in straight lines of wire and fiber. As the amount of nodes increased, the maps grew busier and the straight lines multiplied, softening into wide curves to accommodate their multitudes. Finally, the geographical background disappeared from Internet maps, and the network itself went sovereign. Today, a map of the Internet is a tensile, crazy, fractal thing; it resembles a beating heart, a web of synapses, a supernova.

On top of all this sits the World Wide Web, a network of interconnected visual pages built in a shared language called HTML, or Hypertext Markup Language. "Hypertext" is not a word we use frequently today, but much of the Web is built from these hypertext documents: structured pages of text, images, and video dotted with clickable links connecting individual points to one another. Those connections don't just influence how we *navigate* the Web—Google built its empire on a search engine that brought up Web pages with the highest number and quality of hypertext links—but how we communicate with one another, and ultimately how we understand the world.

In a way, it's fundamental. The Talmud is a hypertext, with layers of annotations arranged in concentric rectangles around a theological heart. Any text referencing another is considered a form of hypertext: sequels, which begin where the last page of the previous book left off; footnotes; endnotes; marginalia; and parenthetical asides. Sprawling, self-referential novels like *Ulysses* or *Finnegan's Wake* are like flattened hypertexts, and scholars love to cite "The Garden of Forking Paths," a short story by the Argentine writer Jorge Luis Borges, as the height of precomputer hypertext. "This web of time," Borges wrote, "the strands of which approach one another, bifurcate, intersect or ignore each other through the centuries—embraces every possibility." He may have loved the World Wide Web.

The Web as we know it isn't modeled on Borges, Joyce, or the Talmud. The most famous hypertext pioneers are men—Doug Engelbart, Jake Feinler's mentor at Stanford, incorporated hypertext into his oNLine System, and Ted Nelson, a Bay Area counterculture hero, coined the word and has championed utopian hypertext ideas for decades—but the Web appeared on the scene only *after* hypertext principles and conventions had been explored for nearly a decade by brilliant female researchers and computer scientists. They were the architects of the hypertext systems that time forgot, systems with names like Intermedia, Microcosm, Aquanet, NoteCards, and VIKI, the earliest ontological frameworks of the information age. Hypertext is, in many ways, the practice of transforming pure data into knowledge. And like programming a generation before, it was where the women were.

MICROCOSM

To understand hypertext, I've turned to one of the brightest computer scientists in the world. Dame Wendy Hall is a garrulous, strawberry blonde Brit with a disarmingly warm manner and a busy calendar. We're talking over Skype, nine hours apart. Wendy, who was appointed Dame Commander of the Order of the British Empire—the female equivalent of being knighted—in 2009 for her contributions to computer science, is in a hotel room in London, dressed for dinner. I'm in my pajamas, drinking coffee, surrounded by index cards, in my office in Los Angeles. For reasons I don't yet understand, she has chosen this moment to catch me up on medieval European history.

The Battle of Hastings, to be precise. "It's something we learn about in history," Wendy tells me, unsure if news of William the Conqueror's eleventh-century triumph ever made it stateside. William the Conqueror earned his nickname by invading England from its southern coast and defeating the last Anglo-Saxon king, Wendy explains, as I indulge this diversion. A few years later, he decided to take stock of his spoils, the entire Saxon kingdom. He ordered an audit of everything he owned. "Every cow, every sheep, every person, every house, every village, everything," she says. "They went 'round, by hand, counting everything up."

The result was an unusual book, now invaluable to historians, detailing the minutiae of the Saxon world, the only survey of its kind. Because the judgments made by the Norman assessors who compiled it were supposed to be definitive, native English people called it the *Domesday Book,* Middle English for "Doomsday Book." As the British cleric Richard FitzNeal wrote nearly a century later, decisions made in the

Domesday Book, "like those of the Last Judgement, are unalterable." The *Domesday Book* is what led Wendy Hall, in a circuitous way, to her career creating hypertext systems long before the dawn of the Web.

In 1986, as Wendy was beginning her teaching career, the BBC—"that's the British Broadcasting Corporation," she reminds me, gently—celebrated the nine hundredth anniversary of the *Domesday Book* by updating it for the modern world, issuing a new British census on a pair of multimedia video LaserDiscs, then the height of technological sophistication. They called them, of course, the *Domesday Discs*. More than a million people contributed to the project, which became a massive volunteer time capsule encoded in bits and light. "Every school in the country was asked to send in three photos of their area," Wendy explains. Schoolchildren wrote accounts of their day-to-day life; Britons sent in photos of office parks, pubs, and windmills. One child, from the village of Spennymoor, contributed this spot-on prediction of the future:

> Robot limbs will be used when natural limbs are lost.
>
> Computers will take over much of the diagnosis now made by doctors.
>
> Food will be made tastier by artificial means.
>
> Children will learn chiefly by computers.

This crowdsourced survey made up the first of the two *Domesday Discs;* the second was filled with interactive material about British heritage, government, and royalty, including census data and some early virtual reality–like tours of notable sites. When Wendy saw the *Discs* for the first time, they blew her away. It wasn't the information that impressed her as much as the *way* it was displayed. "The ideas were stunning," she tells me. The *Domesday Discs* were interactive, using interconnected links that could be navigated with a pointer, much as we're accustomed to doing on the Web today. For Wendy, moving with ease from first-person reflections on British life to census data and 3-D photo tours was a rich, rewarding, and immersive experience. She'd never seen a computer do anything like it before.

Of course, she'd never been much interested in computers. Although her alma mater, the University of Southampton, was one of the first schools in the United Kingdom to teach computer science, Wendy was a student of pure mathematics. According to her doctoral adviser, Wendy was in those years a "shy and retiring student," working in an area of topology "so obscure that to this day I can't understand the title of the thesis." She learned some programming in a first-year course but found it tedious and impersonal. "I was happy in my world of mathematics, and really didn't see, then, that computers would ever really offer me anything," she told a radio interviewer in 2013. But once she saw the *Domesday Discs,* she overlooked her distaste. Suddenly, she understood what kinds of experiences computers could make possible, and as personal computers began to appear in the United Kingdom, "I began to see the future," she says.

Not everyone saw it as clearly. When Wendy returned to the University of Southampton after a stint teaching mathematics to trainee teachers, she accepted a lecturing position in computer science, but her enthusiasm for multimedia was out of step with the established views of the department. "One professor told me in public once that if I carried on doing this multimedia work, there was no future for me at Southampton or in computer science," she remembers, "because I wasn't writing compilers, or new programming languages, or doing operating systems." Many of her colleagues didn't consider interactive multimedia to be real computer science—it was seen as something fluffy, less serious, far closer to the humanities than to classical programming.

But Wendy couldn't shake the glimpse of the future she had seen: a future where images, texts, and ideas were connected through intuitive screen-based links, and computer screens were approachable to all. In 1989, she left Southampton and took a job at the University of Michigan, where she immersed herself in American tech culture, went to conferences, and finally learned that clickable multimedia on computers was indeed a serious discipline, and that it had a name: the Americans called it "hypertext," or "hypermedia." Although her interests seemed out there to most of her British colleagues, she was right at the cutting edge in the United States. She returned to Southampton with a blinding vision for a new hypertext system. In order to explain it, she takes me back in time—again.

This time, we jump decades instead of centuries. Wendy tells me about the Earl of Mountbatten, a second cousin of Elizabeth II. Mountbatten is something of an avatar for twentieth-century Britain. As last Viceroy of India, he oversaw the country's transition into a modern republic. He captained a naval destroyer during the Second World War, and was appointed by Churchill to Supreme Allied Commander of the Southeast Asian theater, where he oversaw a bloody Burmese campaign under monsoon rains. He met Stalin. He met Emperor Hirohito. However, his prominence in British colonial history made him a target. In the summer of 1979, long after his retirement, as Mountbatten was lobster potting from a wooden boat off the coast of County Sligo, he was assassinated along with his family by the Irish Republican Army. They bombed the boat to pieces, blowing his legs almost clear off.

The Mountbattens lived in Romsey, a market village so old the medieval *Domesday Book* made note of its three water mills. According to the schoolchildren who surveyed the village for the *Domesday Disc* project, among the most salient features of life in Romsey in 1986 were a preponderance of punks ("Their spare time, of which they have a lot, is spent hanging around with friends, sometimes playing space-invader machines, with money from dole"), a beloved fish shop, a Waitrose supermarket, and Broadlands, the Palladian estate where the Earl of Mountbatten entertained royal visitors. Broadlands sits on the River Test, which only ten miles south flows past Southampton to join the briny waters of the English Channel. As it happens, the University of Southampton is known for its archives—which is why, after his violent assassination, the records of the Earl of Mountbatten's considerable life ended up in its library.

At Southampton, the Mountbatten archive joined millions of manuscripts, but it was distinctly modern in comparison to most of the library's holdings. Because Mountbatten's public life transected all the touchstones of twentieth-century media, the library inherited some fifty thousand photographs, speeches recorded on 78 rpm records, and a large collection of film and video. There was no linear sequence to the material, save chronological order, and no hope of fitting it neatly into a database. Ten years after the library's acquisition of the Mountbatten archive, Wendy Hall returned from her hypertext sabbatical in America.

Not long after she'd settled back in, she heard a knock at her office door. Word of her interests had drifted from the computer science department to the library. "The archivist came to see me," Wendy remembers, "and he said, 'Couldn't we do something wonderful? I've got this multimedia archive, it's got pictures, it's got film and it's got sound. Couldn't we put it on a computer and link it all together?' And that was the beginning."

The Mountbatten archive was the perfect test case for a hypertext project: a vast, interrelated collection of documents spanning many different media, subject to as many interpretations as there could be perspectives on the last century of British history. Wendy put together a team, and by Christmas 1989, they had a running demo for a system called Microcosm. It was a remarkable design: just as the World Wide Web would a few years later, Microcosm demonstrated a new, intuitive way of navigating the massive amounts of multimedia information computer memory made accessible. Using multimedia navigation and intelligent links, it made information dynamic, alive, and adapted to the user. In fact, it wasn't like the Web at all. It was better.

Microcosm's core innovation was the way it treated links. Where the Web focuses on connecting documents across a network, Wendy was more interested in the *nature* of those connections, how discrete ideas linked together, and why—what we would today call "metadata." Rather than embedding links in documents, as the Web embeds links in its pages, Microcosm kept links separated, in a database meant to be regularly updated and maintained. This "linkbase" communicated with

Wendy Hall demonstrating Microcosm in her research lab in the Department of Computer Science at the University of Southampton.

documents without leaving a mark on any underlying document, making a link in Microcosm a kind of flexible information overlay, rather than a structural change to the material.

To use one of Wendy's examples, say I'm browsing the Mountbatten archive using her system, Microcosm, circa 1989. I'm interested in Mountbatten's career in India, a two-year period during which he oversaw the country's transition from colonial rule to independent statehood. This history has its recurring characters: his field marshal, the leader of the Indian National Congress, Jawaharlal Nehru, and of course, Mahatma Gandhi, whose name is everywhere in the source material. Say also that within the Microcosm linkbase, an instance of the name "Mahatma Gandhi" has been linked to some multimedia information—a video, perhaps, of a Gandhi speech. Because of the nature of Microcosm links, that connection isn't isolated to a single, underlined, hyperlink-blue instance of those words. Rather, it's connected to the *idea* of Gandhi, following the man wherever his name may turn up, across every document in the system. Further, if I were to bring a new document into

Microcosm, the system would automatically identify any words corresponding to links in the linkbase and update it accordingly. Imagine the analog on the Web we know today: for every name, for every idea, *for every linkable thing,* a single repository of supplementary material, updated by everyone in the world, filtered based on parameters determined by the user.

Links in Microcosm could be tailored to the user's knowledge level and could point to several places in the linkbase at once. Microcosm was even able to dig up new links on the fly by running simple text searches on all the material in the system—a prescient design that anticipated the importance of search in navigating information. This "generic" linking, in concert with linkbases, created a system that could adapt to its users while presenting them with more opportunities to learn. "Links in themselves are a valuable store of knowledge," Wendy explained. "If this knowledge is bound too tightly to the documents, then it cannot be applied to new data." Which is to say: where one instance of a connection might be interesting, multiple instances, expressed laterally, look more like truth. By making space for this generic knowledge, Wendy's system placed value on the association *between* documents, rather than on the documents themselves. To hypertext's small but active community of scholars, this is what the field was all about.

In the years between 1984 and 1991, a flurry of hypertext systems like Microcosm emerged from universities and from research labs at technology companies like Apple, IBM, Xerox, Symbolics, and Sun Microsystems. Each suggested different linking conventions, spatial associations, and levels of micro- and meta-precision over contained corpora of information. If this sounds dry, remember that managing, navigating, and optimizing information is a central pursuit of modern life—we do it fifty times a day before breakfast—and that each one of these systems had the potential to become as important to us as the Web is today.

The young discipline of hypertext was heavily populated with women. Nearly every major team building hypertext systems had women in senior positions, if not at the helm. At Brown University, several women, including Nicole Yankelovich and Karen Catlin, worked on the development of Intermedia, a visionary hypertext system that connected five distinct applications into one "scholar's workstation," and invented, in the process, the "anchor link." Intermedia inspired Apple, which had partially funded the project, to integrate hypermedia concepts into its operating systems. Amy Pearl, from Sun Microsystems, developed Sun's Link Service, an open hypertext system; Janet Walker at Symbolics single-handedly created the Symbolics Document Examiner, the first system to incorporate bookmarks, an idea that eventually made its way into modern Web browsers.

For women interested in the nature and future of computers, hypertext was far more collegial than other areas of computer science, which was at that time seeing a rapid decline in female participation at both the academic and professional level. The reasons for this aren't cut-and-dried, but they reflect some of the same tendencies at play over previous generations. While "the whole foundation of hypertext is collaborative," suggests Intermedia's Nicole Yankelovich, and "collaborative work appeals to women," hypertext was also, like programming before it, an entirely new field, a clean slate upon which women could mark their place. Further, hypertext was open to scholars from outside computer science departments, who emerged from such wide-ranging disciplines as interface design and sociology. What these people shared was a humanistic, user-driven approach. To them, the final product wasn't always software: it was the *effect* software had on people.

But I don't really get that until I start talking to Cathy Marshall.

NOTECARDS

Cathy is a hypertext researcher who spent most of her professional career at Xerox PARC, a Palo Alto think tank founded by the printer company in 1970 to help invent the paperless office of the future. "I have to ask you a little about your process," she says, in our first interview.

Interviewing hypertext researchers entails its share of going under the microscope: many of them have never shaken their professional interest in how people organize their thoughts. The second time Cathy and I talk, I find myself describing to her my office pin board of index cards and serial killer–like yarn threads. She shares her own approach. "If I write something and if it doesn't work, I'll throw out the whole thing and start again," she tells me. "I don't think you lose what you've written. It's still in your head. Over time, what you're doing is changing what's in your mind—what's on paper is just incidental." She makes this comment offhand, but the insight knocks me out. That's what software is, I realize: a *system for changing your mind.*

Cathy grew up in Los Angeles and was one of the first women to attend CalTech, after it went coed in 1970. She was only sixteen, small—under five feet on a good day—and mildly averse to math and science; the summer she matriculated, she was more interested in *The Rocky Horror Picture Show,* J.D. Salinger, and macramé than differential equations. When she asked for help, she remembers one professor who told her she'd be better off as a housewife—"my housewifing skills are even worse than my math skills," she thought at the time—and another who waited until the day she missed class to tell a dirty joke. A fellow student filled her in on the punch line, which she still remembers, something about a whorehouse keeping down the "fucking overhead." It wasn't the joke that bothered her—it's that the professor waited to tell it. "I felt so conspicuous and weird," she says.

When she graduated from CalTech with an English degree, she started working as a systems analyst for a company that did radars and signal processing in Santa Monica. Her office had a view of the ocean, but she wrote utility code for a Prime microcomputer, about as dry a technical job as you could get in those days. She went to work at Xerox PARC in the mid-1980s. Even outside computing circles, PARC was known for its freewheeling, hothouse approach: engineers worked alongside anthropologists on a terraced campus built into a wooded hillside overlooking Silicon Valley, and important meetings were held in a room full of corduroy beanbag chairs, low and cozy enough that nobody would be tempted to jump up and attack anyone else's ideas.

After so many years of feeling like an outsider, Cathy worked hard to become part of Xerox PARC's unorthodox workplace culture. "The thing I loved most about PARC is that it was really multidisciplinary," she remembers. "I think it would be hard to find a place that was like that now. They weren't afraid to hire people that had different kinds of backgrounds." The mix was fun: she'd sometimes get silly with the computer scientists, putting Ivory soap in the office microwave until it made huge piles of snaking froth, and she participated in PARC's Artist-in-Residence Program, which paired artists with technologists to create ambitious new media works. Her partner, Judy Malloy, a poet, would often cut a sideways path across the PARC campus, through a field with horses and under a barbed-wire fence, just to pass by Cathy's office window and wave hello.

NoteCards, the first system Cathy worked on at Xerox PARC, was modeled after the kinds of old-school writing techniques about which we'd soon find ourselves debating the relative merits. The software emulated "the way you wrote papers when you were in junior high: with notecards and file boxes." Using hypertext links, users could chain their cards into complex collections, sequences, and mental maps, modeling their thought processes and making it easier for others to understand their conclusions. NoteCards wasn't a writing tool, and it wasn't an information browser like Microcosm, either. When pressed, Cathy calls it an "idea processor."

Hypertext is to text as the technical grammar of cinema is to celluloid: words on-screen become a dynamic medium through buttons and links, just as jump cuts and editing tricks turn moving images into movies. This grammar can be applied to any kind of text, making hypertext highly useful for everything from browsing the Web, as we do today, to idea processing and writing choose-your-own-adventure fiction. NoteCards was designed for intelligence analysis. Before recommending policy, Cathy figured, intelligence brass might want to examine the underlying argument. "Back then I was naïve," she says, and laughs.

The intelligence community never picked up on hypertext, but NoteCards fit perfectly into the multidisciplinary atmosphere of Xerox PARC, a place where anthropologists, linguists, physicists, and computer scientists worked alongside one another. Installed on all the campus workstations, NoteCards became a vital tool for sharing ideas across disciplines, and its influence flowed beyond PARC's borders. In 1987, Apple released a NoteCards-like application, HyperCard, which came bundled with Apple Macintosh and Apple IIGS computers and became the most popular hypermedia system ever developed before the advent of the World Wide Web. People used it to build databases, write branching novels, and create PowerPoint-like presentation slides. Popular games, like the bestselling CD-ROM *Myst,* were prototyped in HyperCard. Within Apple, it was often used to test out interface design ideas, and some publishers even issued magazines as HyperCard "stacks."

Nineteen eighty-seven was a banner year for hypertext, as it happens. Beyond the release of HyperCard, it marked the first academic hypertext conference, Hypertext '87, in Chapel Hill, North Carolina. Academic conferences of this type can forge intellectual communities out of atomized researchers, and this is what happened in North Carolina. Twice as many delegates as expected showed up, leading one attendee to observe a "rueful sense that this was the last time any hypertext gathering will be of manageable size." It was a heady mix, unusual for a technical conference, due largely to hypertext's many applications in the humanities: computer scientists rubbed elbows with classicists, professors with entrepreneurs. "The hypertext conferences were lovely, wonderful in those days," Wendy Hall tells me. "We had what I call the literati there, the poets and the writers. I think that's why it attracted more women."

"Computer science has always marginalized people that are interested in users," explains Cathy, but they found common ground in hypertext research, which was really the study of how people use computers to organize thoughts and data. Those who attended Hypertext '87 came home emboldened by the realization that hypertext wasn't an esoteric interest pursued only by a few fanatics but rather a true movement—one to which tech giants like Apple had clearly been paying attention. "There were little islands of ideas, when we started," Cathy remembers, but as the community coalesced, scholars like Cathy and

Wendy began to think of their vastly different systems as part of a whole. The hypertext systems to come would influence one another in manifold ways, progressively refining the ideas that undergird our century's most transformative information technology.

Part of being interested in users is paying attention to *how* they use software once it's in their hands. Surveying a group of Xerox PARC scholars working in NoteCards, Cathy's colleagues found that although each person "inhabited" the system differently, most used it to plot the big picture: organizing and structuring, sketching outlines, and maintaining references. Building connections and viewing them globally helped writers work through their arguments and ideas, and since NoteCards allowed multiple arrangements to exist in parallel, writers could explore various interpretations before settling. Cathy called this kind of work "knowledge-structuring," and it would dominate her subsequent research. The children of NoteCards—Aquanet, a system named after a hairspray because it held knowledge in place, and VIKI, the first spatial hypertext system—allowed users to organize ideas spatially from the outset, creating graphical schemas for how things fit together. Studying philosophy and logic, and consulting with the anthropologists and social scientists at Xerox PARC, Cathy learned how interpreting material and developing a position is often a process of abstract associations "difficult to articulate within the bounds of language, no matter how informal." Her hypertext systems were meant to empower kinesthetic thinking, the process of moving things around and trying them out akin to "wiggling molecular models in space or moving a jigsaw puzzle piece into different orientations."

All of this might sound bogglingly abstract and strange. Why spend so much time arranging boxes on a screen? But even in the physical world, the piles and clusters we make reflect our thinking: I'm reminded of Jake Feinler's desk at the NIC, covered in precious piles of paper, and of my own desk at home, with its mountains of dog-eared books, notepads, and printouts. Their proximity to one another, and their distance from arms' reach, suggest thematic connections and conceptual closeness to my thought process. In an influential paper of the hypertext era, Alison

Kidd, a researcher at Hewlett-Packard, called such piles "spatial holding patterns," suggesting that they play an important role in "creating, exploring and changing structures which can inform us in novel ways."

Cathy's hypertext systems shifted all these mental patterns on-screen and integrated them into larger writing and argument-building environments, presaging the ways in which we'd all soon find ourselves working on computers, with ever-expanding tabs, documents, and apps organized to suit our particular thought processes. They also demonstrate just how complex and nuanced hypertext can be, when the technology is explored to its fullest potential: it supports not just links but entire mental maps, systems that model—and more important, change—our minds.

This is the kind of thinking that prompts Cathy, thirty-odd years later, to tell me that what's on paper is incidental. That the only important thing is what stays in your head. If my documents, strewn on my desk or clustered as icons on a screen, appear inscrutable to an outside observer, that's no flaw in my system. They *should* be meaningless, because they're only the remnants of a transformation process, like a sheaf of molted skin. The real technology is the user.

That means me. And you.

HYPERTEXT '91

It all came to a head at Hypertext '91.

The conference was held that year in San Antonio, Texas. North Carolina had indeed been the first and last time the hypertext community would be a manageable size—in the four years since Hypertext '87, it had exploded, and academics, writers, engineers, and developers from around the world converged in Texas for the occasion. Wendy Hall came from England to demonstrate the latest build of Microcosm. The conference floor, a hotel reception area lined with rows of tables, was clustered with representatives from dozens of hypertext projects with names like AnswerBook and LinkWorks. Several tables down from Wendy Hall sat another British computer scientist, Tim Berners-Lee.

He'd had his conference paper rejected, but he'd come to San Antonio anyway, to show off a new system to the hypertext crowd.

He'd brought Robert Caillau, a colleague from CERN, the European Organization for Nuclear Research. The pair was demonstrating a distributed hypertext system Berners-Lee had built to make sharing data on networked computers across their massive Swiss campus a little easier. To anyone who saw it in 1991, it would have looked something like NoteCards or Apple's HyperCard: small graphical "pages" connected by links. The major difference was that these pages didn't all live on the same computer; Berners-Lee and Caillau, in the hopes of making data accessible to physicists outside of CERN, had built their hypertext system on the backbone of the academic Internet. They called it the World Wide Web.

To demonstrate the World Wide Web, Berners-Lee and Caillau brought their own computer with them on the plane from Geneva: a ten-thousand-dollar jet-black NeXT cube, at the time the only machine capable of running Berners-Lee's graphical World Wide Web browser. Still, the hypertext community wasn't impressed. "He said you needed an Internet connection," remembers Cathy Marshall, "and I thought, 'Well, that's expensive.'" Wendy Hall took a break from her own demo to try the Web on the conference floor. "I was looking at it," she remembers, laughing ruefully, "and I'm thinking, 'These links, they're embedded in the documents, and they're only going one way—this is really too simplistic.'"

They were right. It was expensive. Although Stanford had established the first stateside Web server only three days before, the hotel in San Antonio wasn't fronting for an Internet connection, so Berners-Lee and Caillau were forced to demo a dummy version of the Web saved on optical disk. And it *was* too simplistic. Compared with the other systems on display, the Web's version of hypertext was years behind. Links on the World Wide Web went in only one direction, to a single destination, and they were contextual—tethered to their point of origin—rather than generic, like Wendy's Microcosm links. Instead of employing a linkbase that could update documents automatically when links were moved or deleted, the Web embedded links in documents themselves. "That was all considered counter to what we were doing at the time," Cathy adds. "It was kind of like, well: we know better than to do that."

Because the demonstrations at Hypertext '91 were scheduled after all the day's lectures and discussions, many delegates skipped the session entirely. Reporting after the fact, a contributor to the field's journal of record, the *ACM SIGCHI Bulletin,* noted that she had "little energy left to see and understand the demos, let alone try to conduct an intelligent conversation about them." Of the twenty-one demos on display, she managed to examine only six, one of which was the World Wide Web. In what amounts to little more than a footnote in her trip report, she calls it a "hypertext-like interface" intended for the "High Energy Physics community."

Nothing in the proceedings of Hypertext '91 suggests how quickly the World Wide Web would come to dominate the lives of people all around the world—and indeed, to alter the course of human history. In San Antonio, it was just one of many systems on display, and far from the most sophisticated. It certainly didn't help that the techno-social activities of Hypertext '91 featured a tequila fountain in the courtyard outside the hotel. The very moment the World Wide Web was making its American debut, everybody was outside drinking margaritas.

But the Web suffered little from its snub in Texas. By Hypertext '93, more than half the demos on display were Web based, and at the European conference on hypertext in '94, Berners-Lee delivered the keynote address. During that brief window, hypertext and the Web managed an uneasy coexistence. Cathy Marshall proposed that hypertext systems could serve as desktop work spaces for information gathered online, and HyperCard wasn't pulled from the shelves until 2004. Wendy Hall, who would weather the transition to the Web more successfully than many of her peers, updated Microcosm to include a Web viewer, and she designed versions of her beloved linkbases that could be shared over a distributed network.

Today, we mostly think of hypertext as being something related to the Web, rather than of the Web as a technically inferior manifestation of hypertext principles. The Web is hypertext's killer app, just as e-mail was the Internet's killer app—but its success hit the hypertext community hard. "I'm not sure exactly how to describe it," Cathy Marshall tells me. "All of a sudden you were the outsider, when you'd been the insider." At the first World Wide Web conference in 1994, Wendy Hall noticed that many delegates thought the Web was the first hypertext system, and she was stunned to read a paper reinventing her generic link ideas from scratch. By 1997, the two fields were so divergent that the Hypertext and World Wide Web conferences were scheduled for the exact same week.

To this day, the World Wide Web suffers from problems that systems like Microcosm solved decades ago. Because Web links are entirely dependent on their context, they're almost impossible to maintain. If a Web site is moved, deleted, or hidden behind a paywall, every link that pointed to it becomes meaningless, dangling like an anchor line cut loose from a ship. This should be familiar to anyone who has spent five minutes browsing the Web: according to a 2013 study, the median life span of a Web page is 9.3 years, a rate of obsolescence that sows rotten links throughout the network over time. We all regularly experience these dead ends, which are called 404 Errors. The document you're looking for, they tell us, simply cannot be found.

The hypertext researchers who demoed the World Wide Web back in San Antonio assumed this issue would be the system's undoing. After all, what good is a hypertext system if the links don't even work? Further, the Web isn't constructive. In all of the major pre-Web hypertext systems—Microcosm, NoteCards, Aquanet, VIKI, and Intermedia—creating links was just as important as clicking them. The point was for users to build their own paths through the material, a creative process of forging associative trails that could be shared with others. The Web, however, is a passive medium, a highway we wander without leaving much of a trace.

The World Wide Web may not have been powerful enough for academics, but a lightweight, user-friendly tool is often more likely to take off than a vastly more powerful one. And while linkbases and constructive hypertext were easily maintained in relatively contained research and classroom environments, or on small networks of computers all running the same operating system, they would have quickly become unmanageable on a global scale. Today, we accept 404 Errors as the cost of doing business, and the Web runs the world.

MULTICOSM

The second time I talk to Wendy Hall, she's finishing up a long day with the department she now chairs at Southampton, the Web Science Institute. As I reach her on Skype, she's just saying good-bye to the last students trailing out of the conference room where they've been meeting. "Claire's writing a book about me," she says, laughing, to someone I can't see, gesturing at my head on the screen. "Or people like me, anyway."

Wendy will be the first to tell you that she's a very social person. She loves to make connections with people, and between them. When she talks, she does so in long, unselfconscious streams, jumping from one big, seemingly unrelated idea to another on her own invisible tracks—the mark of a true hypertext researcher. She loves science fiction and asks me repeatedly if I've read Douglas Adams's *Hitchhiker's Guide to the Galaxy* or Isaac Asimov's Foundation series. These novels contain her go-to analogies: the World Wide Web, she says, is an experiment on the whole world, just like the white mice in *Hitchhiker's Guide*, who run through mazes to test the scientists, and trying to understand the Web is like studying Isaac Asimov's "psychohistory," a mathematics of social complexity that can predict the rise and fall of galaxies.

Back in 1991, after drinking her fill of courtyard margaritas in San Antonio, Wendy went back to Southampton to continue developing her multimedia hypertext system, Microcosm. To survive, it needed to adapt to changing times. This it did admirably: for every new form of media, Wendy and her team developed new Microcosm "viewers," windows through which its users could draw material into their growing personal linkbases. There were Microcosm digital video viewers for video LaserDiscs, viewers for animation, sound, and 3-D models, and viewers for competing hypertext systems. After San Antonio, however, Wendy was careful to add one more: a viewer for the World Wide Web.

The Microcosm Web viewer served as a hypertext replacement for the standard Web browser. Where browsers like Mosaic—and later Netscape and Internet Explorer—were read-only, Microcosm users could, using their Web viewer, select text from anywhere on the Web to use as the starting point or destination of their own hypertext threads, linking to other Web pages, multimedia documents, and their personal Microcosm linkbases. This all seemed to Wendy to be the most natural thing in the world. Many in the hypertext community balked at the Web's brutal simplicity, but Wendy corrected for it, layering her robust—and proprietary—system on top of the more skeletal world of interlinked Web pages. "I saw the Web through a Microcosm viewer," she explains. "Of course, Tim saw it completely the other way around."

Like many hypertext researchers, Wendy had every reason to assume that her system could happily coexist with the Web. After all, Microcosm worked better. It didn't suffer from dead links, and where the Web connected Web pages only to one another, Microcosm connected word-processor documents, spreadsheets, videos, images, and CAD files, like a micro-Internet interlinking *everything* on the desktop. "You could follow links to many different destinations," she says. "You could have one-to-one, one-to-many, many-to-many links. And you could reverse it all." Because Microcosm links were stored in a database rather than embedded into documents, the system could generate new connections on the fly, tailoring itself to individual users' browsing habits. "I was still thinking of the Web as *one* of the systems we would use," Wendy says.

What she didn't anticipate is the network effect. Because the Web was built on top of the Internet, and because it was free, early adopters quickly gave way to more mainstream users, and the more people got on the Web, the more interesting it became to their friends and family, and so on, very quickly making it dominant. Meanwhile, the Microcosm team released new stopgaps. They streamlined the system's key concepts—its generic links and linkbases—into a Web browser add-on called the Distributed Link Service, which made any browser into a kind of Microcosm Lite, applying generic linking principles to the interchange between client and server. Effectively, this allowed Web users to interrogate material on the Web regardless of whether an explicit link was there. "By enhancing the Web with Microcosm's link services, WWW readers would be freed from the tyranny of the button," Wendy's team wrote in a 1994 paper.

The tyranny of the button, however, prevailed. People navigating the Web for the first time, few of whom had any experience with the kind of hypertext people like Wendy and Cathy created, were perfectly happy to click buttons, roaming with no clear destination along the Web's labyrinthine paths. Such curiosity-driven *dérives* were, in fact, part of the Web's early appeal. When people hit a dead end, they went back and tried a different link. The maze was a mess, but it was worth running.

Had Wendy's team at Southampton concentrated its energies differently, it's very possible that Microcosm could have been the first graphical Web browser to really take off. But that would have entailed making the software as free to use as the Web itself, and Wendy's team had its sights on commercializing their efforts. In 1994, they established Multicosm, Ltd.: if one microcosm is a window on the world, their company would produce many. The timing couldn't have been worse. "People used to say," she remembers, and laughs, "'I think what you're doing is wonderful, but this Web thing is free, so we're gonna try that first.'"

Fortunately, Wendy never abandoned university life. Running an expanding department at Southampton, she remained in contact with the growing Web development community, and after working closely with Tim Berners-Lee to develop the Microcosm Web viewer and the Distributed Link Service, she became a sustained presence on the early Web scene. In 1994, she helped to organize the first Web conference but still wasn't confident that the Web was the end-all solution. In a 1997 lecture at Southampton, she minced no words. "The Web has shown us that global hypertext is possible, but it has also shown us that it is easier to put rubbish on the Net than anything of real and lasting value," she said. How right she was.

There is a coda to this story, however. Microcosm's ideas may not have been implemented on a global scale in their day, but their prescience is undeniable. The way Wendy's system created links dynamically, based on the context of the information being linked, was a form of what we now call metadata. "We are now, twenty-seven years after the Web, living in a world that is driven by data," Wendy reminds me. *How* and *why* that data are linked is becoming increasingly important, especially as we teach machines to interpret connections for us—in order for artificial intelligence to understand the Web, it will need an additional layer of machine-readable information on top of our documents, a kind of meta-Web that proponents call the Semantic Web. While humans might understand connections intuitively, and are willing to ignore when links rot or lead nowhere, computers require more consistent information about the source, the destination, and the meaning of every link. "That was the core of Microcosm," Wendy says. When she began to participate in building the Semantic Web in the 2000s, it was "so exciting because I could see all my original research ideas coming to life in the Web world. We still couldn't do all the things we could do in Microcosm in the '90s, but we could see how effective our linkbases were."

In the end, however, the system is immaterial to her. It's the connections she's after: the magnificent complexity of human society and thought, all influencing one another in the unfolding of history. "There's lots of different ways that we could have implemented a global hypertext system," Wendy says. "The Web won—for now. But it feels like this is an experiment that has involved the entire world. Have you read *Hitchhiker's Guide to the Galaxy*?"

PART THREE

The Early True Believers

Chapter Eleven

MISS OUTER BORO

The Internet exists at the confluence of culture, code, and infrastructure. As the technology historian Janet Abbate writes, "Communications media often seem to dematerialize technology, presenting themselves to the user as systems that transmit ideas rather than electrons." This makes the boundary between users and producers, and between software and hardware, so porous as to be effectively permeable. As the story of hypertext shows, technology alone isn't enough to change the world—it has to be implemented in an accessible way and adopted by a community of users who feel enough ownership over it to invent new applications far beyond the imagination of its architects. To make successful links, in short, we need things worth linking.

Even as it eclipsed the more complex hypertext systems preceding it, the Web didn't transform the world on its own. A generation of smart, creative, nonacademic, and nontechnical users had to come along first to forge the connections that would make up the vastly interdependent network we love—or love to hate—today. To fill the container with content, these users would need to be intensely familiar with both computers and culture. They'd need to know how to build and connect. They'd need to understand how something that organizes information can also inspire thought. Fortunately, there were plenty of people like that on Echo. One of them called herself Miss Outer Boro.

Marisa Bowe was a teenager in the 1970s when her dad brought home a wood-paneled box with a square plasma gas screen that glowed faintly orange. He installed it in the basement of the family house in suburban Minneapolis, near Lake Minnetonka, where they would speedboat in the summer. It wasn't a personal computer—it would be several decades before those turned up in suburban basements nationwide. Rather, it was a terminal, mute until the home phone, coupled into a cradle fit for the purpose, dialed the network of supercomputers that brought it to life.

Marisa's father was a PR executive at the Control Data Corporation, an old-school mainframe computer company. In the 1960s, they produced the fastest supercomputers in the world. But by the time Marisa was a teenager, her father's company was placing a risky bet on computer-assisted education, wagering that teachers might someday be replaced with electronic learning systems running specialized software. They called this Programmed Logic for Automatic Teaching Operations, or PLATO. The software was built on their mainframes.

The PLATO model was to offer distributed low-cost education, with a number of centrally hosted teaching programs accessed remotely by students at glowing orange terminals like the one in Marisa's basement. Although PLATO terminals would eventually be installed in universities and schools from Illinois to Cape Town, very few people would ever have one at home. Not that Marisa took advantage of her unique situation to learn anything. PLATO offered lessons in everything from arithmetic to Hebrew, but she used it to talk to boys.

"I didn't have any interest in programming," she tells me, "but I discovered there was live chat. So I started up a flirtation with my father's boss's son. You would sort of be able to see who was online—there were very few people, ever, online. And here I was, a teenage girl, and they were all boys, but I was shy in real life. I think people asked me what

I was wearing; it didn't get dirty or anything like that, but it was very flirty, and it was really fun."

Wherever PLATO terminals were installed, students were chatting, posting messages about "science fiction, women's rights, football, the defense budget, rock 'n' roll." PLATO had a primitive version of e-mail called Personal Notes, public "Group Notes" that served as bulletin boards, and one-on-one chat, called TERMType. The PLATO community piloted all the tropes that would become common in online communities—including men impersonating women—and like the Community Memory terminals emerging simultaneously in the Bay Area, it had its own wonky grassroots counterculture.

Marisa was, in PLATO slang, a *zbrat*, a kid using her father's log-in to mingle on the nascent network. She doesn't remember ever seeing another female zbrat, but she made her share of online friends. Beyond the flirtation with the boss's son, she chatted with college students in Boulder, Colorado, and in Champaign-Urbana, at the University of Illinois, where the PLATO software was being developed.

PLATO users shared recipes, gave each other love advice, used emoticons, and played fantasy role-playing games. For a suburban kid in the Midwest, all of this was totally unexpected, as though a household appliance had suddenly become a window into another dimension. PLATO altered the course of Marisa's life. When most people her age still thought of computers as monolithic calculators in university basements, she understood that once connected, they became social machines. It would take a decade for the rest of the world to catch up—and by the time it did, Marisa had a head start.

As Marisa grew older, she couldn't quite shake her fascination with words on screens. Trying LSD for the first time, she laid on her back in the yard, watching her "mother-of-pearl" cigarette smoke coil beneath the stars, which seemed to her "like a TV set, moving around and forming words which I couldn't read." When she moved to New York in 1985—during the "Donald and Ivana Trump, merger-and-acquisition, junk-bond boom-time"—she discovered BBS culture, and The WELL.

"Oh my God," she thought. "This is just like PLATO but with interesting people!" She was eager to relive her adolescent flirtations, but The WELL didn't do it for her.

On today's Web, geographical distance doesn't count for much, beyond tonal differences—e-mails from abroad arrive at seemingly strange hours, and East Coast late-night Twitter dead-ends into the West Coast morning feed. But in the dial-up days, a significant technological and cultural boundary divided the East and West Coast Internet. These digital Rockies presented Marisa with two insurmountable passes: dialing into The WELL was a long-distance phone call, with the accompanying costs. And once she dialed in, there would be Deadheads. "I'm *allergic* to the Grateful Dead," she confessed in a 2011 interview.

Thankfully, another tech-savvy New Yorker named Stacy Horn had just had an identical crisis. After Marisa Bowe gave up on The WELL, she discovered Echo—a local call—where Jerry Garcia's name was verboten. The way she tells the story, it was love at first keystroke: "I logged on to Echo and four years later I looked up." The diverse community and the alternative it presented from traditional media was seductive and novel. She loved "the idea that you could converse and get opinions from people who weren't, like, the twelve guys from Harvard who ran the *New Yorker, Harper's,* and the *Atlantic*."

Marisa lived in Williamsburg, Brooklyn, so she gave herself the username Miss Outer Boro, as in "borough." As MOB, she posted constantly, to the point where her real-life friends started to worry. "They felt sorry for me," she tells me now, with a gleeful cackle, Skyping from her Williamsburg apartment on a hot summer day almost exactly twenty years later. "They felt it was such a losery thing to do." Little did they know that Miss Outer Boro had a cult following: Marisa's natural instinct for online conversation had made her an instant Echo favorite. Where she was shy in real life, she was bold and irrepressible online. "She makes me spit out my coffee on my keyboard more than any other person on Echo," Stacy Horn wrote.

Stacy appreciated Marisa's presence on Echo so much that she begged her to host the Culture conference. Marisa knew how to draw worthwhile conversation out of people. Although getting a good thread going online could be like pulling teeth, she had the constitution for it: a combination of patience and magnetism, with a quick wit and fast-girl charisma that popped from the screen. "She knows she's smart, powerful and beautiful," wrote Stacy, "part tart"—those formative years cyberflirting on PLATO had served her well—"part total queen." She became the conference manager, the host of hosts, and the object of plenty of one-sided ardor. "Half of Echo was dying of love for her for the longest time." On Echo, Marisa was the ultimate arbiter of cool. The writer Clay Shirky called her "the Henry James of the Alley." In a sense, she was one of the earliest online influencers, so popular that she sometimes felt uncomfortable turning up for Echo's in-person events. "It was like a mini-celebrity experience," she recalls.

But although she was a big deal on the Internet, her real life wasn't always so glamorous. She spent a lot of time holed up in her apartment, nose to a screen of endless text, its white words on blue scrolling like clouds through the sky. To pay rent, she took temp word-processing jobs. It hadn't yet occurred to her that she might be able to combine the two.

THE BIGGEST BITCH IN SILICON ALLEY

In 1994, a new personality tore across Echo. Her real name was Jaime Levy, but in a nihilistic nod to Kurt Cobain's suicide, she posted online as "Kurt's Brains." An inveterate Nirvana fan, she looked the part: her peroxide-blonde hair was usually gelled into upright whorls, and she was rarely photographed without a skateboard, a hand-rolled cigarette, and an oversized flannel shirt.

Jaime grew up a latchkey kid in the San Fernando Valley, freely roaming the haze-tinted sprawl. While her brother stayed indoors to mess around with the family Commodore 64, she preferred the punk rock scene, then at its cultural apex in late 1970s Los Angeles. Computers didn't appeal to her; the command-line language her brother was always banging onto his keyboard didn't look creative at all. In her first year of film school in San Francisco, a boyfriend showed her how to make animations on an Amiga, the successor to her brother's Commodore 64. It clicked: computers are punk. She started adding computer graphics to her experimental films.

By twenty-one she was already too big for her britches in San Francisco. "The video art scene was overrun with people who didn't care about money," she tells me now, at fifty. "I knew that I needed to be viable. I didn't want to be a starving artist." She turned in her thesis film, an effects-laden skate video, and bought a plane ticket to New York City, having decided that NYU's Interactive Telecommunications Program (ITP) was for her. Some of her favorite filmmakers—Jim Jarmusch, Spike Lee—had graduated from NYU, and it seemed like an interesting time to be in New York. There was video art being screened at The Kitchen, and this interesting band, Sonic Youth, was playing gigs in the city. In typical style, she marched into the Tisch Building at NYU without an appointment.

The move was part bombast, part desperation: she didn't think she could have gotten an interview if she tried. She wandered around the fourth floor until someone led her to the office of Red Burns, the program's venerable chair, a bristly redheaded matriarch some called the Godmother of Silicon Alley. Jaime was intimidated, but she had only two days in the city, so she told Burns she wanted to tell stories using computers. Later, she'd joke that everyone at ITP was there "on a Citibank scholarship to design ATM machines." Burns took a shine to Jaime, the kid with a skateboard and an Amiga whom she may have seen as a swift kick to the program. "I think she saw the opportunity to attract younger people who wanted to push technology in new ways for art, for entertainment, for publishing, for everything." Burns gave Jaime a full ride.

Jaime spent her years at NYU experimenting with interactive media. For her master's thesis, she combined the do-it-yourself ethos of punk with the emerging possibilities of desktop publishing, producing an electronic magazine, *Cyber Rag,* on floppy disk. With color-printed labels Krazy Glued onto each disk, *Cyber Rag* looked the part of a punk

Jaime Levy's electronic magazines

rock fanzine. Loaded onto a consumer Mac, Jaime's stories came to life with images pilfered from the *Village Voice*, the *Whole Earth Review*, *Mondo 2000*, and *Newsweek* collaged together on-screen as though they'd been xeroxed by hand. *Cyber Rag* was programmed in Apple Hyper-Card, with graphics drawn in MacPaint. Along with her animations, she added edgy interactive games (in one, you chase Manuel Noriega around Panama), hacker how-tos, and catty musings about hippies, sneaking into computer trade shows, and cyberspace. Before the first graphical Web browsers brought hypertext to the masses, Jaime was publishing disks she thought would replace print magazines. They were rewrite-able, after all. "If you hate it, take the files off, throw 'em away, put your own files on it," she told a reporter in 1993.

After grad school, Jaime moved back to Los Angeles and renamed her magazine *Electronic Hollywood*. When the city was upended by the 1992 L.A. uprising, she couldn't leave her Koreatown apartment for four days. She climbed up to her building's roof with a camcorder and filmed the streets. There were fires everywhere. "We called it Smell-o-vision," she said, "because you could run back downstairs and turn on the TV and there would be smoke coming inside the building." She added those images to *Electronic Hollywood*, alongside an editorial about the experience. "I feel like a survivor of a post-Reagan morning after," it reads, as a clanging industrial sample plays on loop. "I can't even walk to buy beer anymore. Los Angeles, love it or loot it!"

In filmed interviews from this time, Jaime's SoCal affect—to this day, she speaks in a deep valley brogue—belies a very real intensity. In one profile produced by the Los Angeles public television affiliate KCET, she spins her office chair around in a room plastered with concert flyers, describing the electronic magazines offhandedly as "my, sorta, digital graffiti." But despite her show of Gen-X disinterest, Jaime was at the absolute vanguard of electronic publishing. Nobody had ever produced anything like *Cyber Rag* or *Electronic Hollywood*. There had been some interactive HyperCard stacks, downloadable from BBSs, and a few interactive art disks for Commodore Amiga. But Jaime's disks, packaged on floppy, were accessible to anyone with a Mac, and with their hypertext links and interactive animations, they functioned exactly like Web sites—long before the Web existed.

Although she had a day job as a typesetter, she distributed *Electronic Hollywood* to indie book and record stores, where they routinely sold out. The novelty got her national media attention, which she leveraged into mail-order sales. After her magazines were featured in an issue of *Mondo 2000,* the cyberculture's magazine of record, she was flooded with orders and fan mail. Although she made only five issues, she sold more than six thousand copies at six bucks a pop—not bad for disks that cost less than fifty cents to produce.

When Jaime finally moved back to New York, she became Silicon Alley's first real celebrity, a poster girl for a new generation of twenty-something media titans ready to reboot the world. She was the first truly iconoclastic, magazine-profile-ready female face of the emerging digital culture. It helped that she photographed well: surly faced in polyvinyl pants and boots for a 1996 *Esquire* story about "grrrls" who "just wanna be wired," or in her omnipresent grungy flannel and tube socks, clutching a skateboard, in a *Newsweek* "Who's Who." The Net was big news, and she made a compelling figurehead: young, edgy, with attitude for miles. "I was the Kurt Cobain of the Internet," she told the *Village Voice* at the tail end of the decade, only half kidding.

"Jaime really knew how to present herself," remembers Marisa Bowe. "She would go to Mac conventions and diss them in that little zine voice. Zine kids are always saying, 'We hate these bands and we

"I don't feel like me being a chick hurt me at all. It helped because there were so few of us, and I got all the attention because I was the crazy one."

hate those bands.' She was doing the same thing, but about computers, and on a floppy disk with her own music, and I thought it was just brilliant." She was a hacker through and through: in her loft, a television screen doubled as her Mac monitor, and a flashing red light, the kind usually reserved for the hearing-impaired, signaled incoming calls over the din of her stereo. Instead of treating the computer like a precious object, Jaime dragged it by the power cord, banging every curb along the mean streets of L.A. and New York. The first issue of *Electronic Hollywood* opened with a tongue-in-cheek greeting that pretty much sums up her style at the time: "If you have never seen an electronic magazine before," it read, "then I hope this trips you out."

Anyone who was anyone in the emerging tech counterculture had copies of *Cyber Rag* and *Electronic Hollywood*—even Billy Idol, the British rock star, was a fan. He came across *Electronic Hollywood* at the height of his own cyberpunk phase, as he was working on a record about sex, drugs, and computers called, appropriately, *Cyberpunk*. Idol loved Jaime's floppies and decided he needed one for his album. They worked out a deal: for five thousand dollars, Jaime would make Billy Idol an interactive floppy disk of his own. It would be just like hers, with his lyrics in place of her rants and band reviews, and would come packaged in a folding cardboard sleeve alongside his CD. Her first corporate gig—and the world's first interactive press kit. "He just really wanted to party with me," Jaime remembers. One night, at Manhattan's infamous Club Fuck, they got so wasted that Billy threw her across the table and broke her arm. "I was supposed to be animating his disk. I had to, like, do the whole thing with my non-mouse hand."

Idol's album flopped, but Jaime kept her rep as savvy netizen. She got a straight job, commuting to White Plains to do "bonehead interface design" at IBM—in her baggy plaid shirts, she was often mistaken for a janitor—and was tapped by Viacom to make an interactive press kit for Aerosmith on CD-ROM. But she was too young and too punk to go totally corporate, and she was getting bored with physical media. "Who's going to buy this shit? Who cares about these CD-ROMS?" she wondered. A coworker at IBM showed her Mosaic, the first browser for navigating the World Wide Web. She experienced it as a conversion: her electronic magazines were Web sites, before Web sites existed, built exactly the same way, with hypertext navigation and "pages" of sound, video, and text. "Once the browser came out, I was like, 'I'm not making fixed-format anymore. I'm learning HTML and that was it.'" She quit her day job.

Around that time, Jaime started hosting parties in her Avenue A loft. She called them "CyberSlacker" parties, updating the Gen-X honorific for the wired generation. CyberSlacker was the first of the Silicon Alley parties. Within a few years, the flush of dot-com money in New York would turn the Flatiron district into a bacchanal of rare proportions. Multimillion-dollar startups with names like Razorfish and DoubleClick burned their IPO money on go-go dancers and vodka pyramids as their CEOs became celebrities, their risqué party pictures leaping from the *Silicon Alley Reporter* to *Page 6*. The most infamous of the Alley parties were funded by an early Web-streaming company, Pseudo, whose

CEO, Josh Harris, was notorious for bringing together tech socialites, hungry entrepreneurs, artists, musicians, and New York club kids in increasingly extravagant environments. On the eve of the millennium, Harris burned more than a million dollars on a monthlong experiment in communal living. Over two adjoining warehouses, "Citizens" in matching uniforms lived, ate, drank, had sex, and shot guns under the gaze of hundreds of streaming digital cameras until they were shut down by the NYPD on New Year's Day 2000. The cops thought they were a Y2K cult. Maybe they were.

"Josh Harris always *says* he didn't copy me," says Jaime, who would never have his resources, "but it's impossible that he didn't." Her parties may not have been as outrageous, but they had a bigger effect: part rave, part hackathon, CyberSlacker sparked the tinder of a uniquely New York tech scene, which was defined by a preponderance, as one historian puts it, of "principled slackers, arty punk rockers, and deconstructionists from 'good' families." Many of these saw the Web for the first time in Jaime's loft, on a Mac II her hacker friend Phiber Optik—Echo's sometime tech support—set up with a 28.8K Internet connection. As avant-garde guitarist Elliot Sharp performed live, and another friend, DJ Spooky, played house tracks, Jaime's guests gathered around the Mac's small screen. At the top of 1994, there were fewer than one thousand Web sites in the world, mostly personal home pages. "That's all that was up there," Jaime remembers. "Like, 'Hi, my name is Lisa and here's my dog.'" From these humble beginnings, however, the right witnesses extrapolated the essential—like Jaime, they saw how irrelevant the Web made everything they'd done before.

These converts would call themselves the "early true believers," counting the year of their arrival online as a mark of status, the way the first punks claimed 1977. Soon after the bubble popped, a *New York* magazine story about Jaime and her peers nailed the time line: "Nineteen ninety-five is cool. Nineteen ninety-six or early 1997 is all right. Anything after that is not. Two thousand makes you a real loser—a suit, a kid just out of college, a fiftyish businessman looking for one last hurrah and another hundred million dollars."

The Web gave the early true believers creative, spiritual purpose and offered them an opportunity to sidestep the media giants that seemed to hold the only key to creative employment in recession-era New York. "It turned us all into apostles," one Alley scenester proclaimed. "It wasn't a money thing; it was, 'Hey, here's this pure channel.'" Then, of course, the money came to town, and—for a short while—they became the giants.

WORD

Jaime was always telling Marisa Bowe she should get into electronic publishing. "I didn't know what she meant," Marisa says. "I mean, I liked electronics and I liked publishing, so I probably would like it." She wasn't alone—beyond Jaime's floppies, and a few early experiments on the Web, publishing online was an unknown quantity.

For Marisa, that would change in 1995, when a software company called Icon CMT tapped Jaime, then at the height of her fame, to be the creative director of a new online magazine called *Word*. The budget was good, or at least more than nothing, which is what Jaime had in the *Cyber Rag* days. The editor in chief, Jonathan Van Meter, was an industry veteran who'd previously helmed *VIBE*. It seemed like he'd let Jaime be Jaime: the snarky, pot-smoking doyenne of the cyberunderground. The team was hers to assemble.

The last thing she wanted was for *Word* to feel trendy, like Billy Idol's failed foray into cyberpunk. She trusted only one person to ensure *Word* would be cool: her friend Marisa Bowe, aka Miss Outer Boro, whose spellbinding effect she'd witnessed firsthand on Echo. When Marisa went to interview for the job, she decided not "to pretend like I'm some super-straight magazine person," she tells me—and "by straight, I mean, gets along well in the corporate world." But she and Van Meter hit it off, and he hired her as *Word*'s managing editor. Having never worked in the magazine world, she had no idea what that meant.

In fact, putting a magazine on the Web at all was a new proposition: *Word* would be among the first. To say there was no business model is an understatement. *Word*'s backers assumed that online publishing would be a subscription business: like traditional magazines, readers would subscribe, and because there'd be no printing or distribution costs, the profit margin would be huge. They'd be selling something ineffable, again and again, without ever running low on inventory—the dot-com era's delusion of choice. Icon thought they'd make millions in a matter of months.

Van Meter dropped out shortly before *Word*'s 1995 launch. Jaime says it's because he didn't know anything about the Web. Marisa suggests that Icon hadn't been straight with him about the editorial budget, which he was accustomed to having in Condé Nast proportions. His absence left Marisa at the helm, technically unqualified but totally right for the job. With her head start on online culture, she "knew how to approach a medium by looking at what it was."

Marisa would later say that while she appreciated what would become *Word*'s primary competition—brainy online magazines like *Salon, Slate,* and *Feed*—she loved the homespun feeling of bulletin board systems and personal home pages much more. Amateur writing, Marisa observed, is the bread and butter of the Internet. In a 1996 roundtable discussion with a handful of other online editors, she confessed to being "hooked on amateur stuff. At best, it's fascinating—passionate, intimate, unpredictable, and disarmingly lacking in artifice."

She brought that sensibility to *Word,* which published, almost exclusively, accounts of everyday people's experiences. The site's most popular section featured weekly testimonials from people in every line of work imaginable, edited down from interviews largely conducted by Marisa's brother, John, a Studs Terkel for the digital village. He went on cross-country road trips to find real people to interview, and *Word* published his transcribed accounts of the lives of prostitutes, hat salesmen, UPS deliverymen, and heavy metal roadies.

Word staff weren't above sharing their own stories, either. The masthead was a showcase of entries from everyone in the office, and Marisa went so far as to publish snippets of her own high school journals, mastering the confessional tone of proto-blogs. Her very first editorial was an abridged life story: speedboating on Lake Minnetonka, her father's ice cubes rattling in the boat's cupholder; being first on the school bus and singing along to "King of the Road" with the acne-riddled driver; hating her "fascist-looking" Brownie uniform. It includes a recipe for beans and a passage in pig latin. *Word* was documentary, not commentary. Marisa felt that what mattered about the Internet—and what was at risk of being lost—was the voice of individual people. "What was fundamentally most fascinating and different about the Internet," Marisa tells me, "was that *people* are the medium that you're working with."

As Marisa honed authentic stories, chasing the human touch that had first hooked her as a teenage zbrat, Jaime Levy focused on more literal interactivity. She described her goals to a reporter soon after the site launched: "In the world of *Word,* we don't just read 'See Spot Run,' we have a three-dimensional encounter, in which Spot not only barks and bites but in which you might want to carry a pooper-scooper."

Like that analogy, the interactive features Jaime designed for *Word* were both startlingly original and characteristically caustic, like *Word*'s chatbot, Fred the Webmate—a pixelated, misanthropic New Yorker you could talk to by typing questions. Fred was damaged, with sexual issues he often alluded to but never addressed. If pressed, he claimed to have been recently laid off by a large new media company. "Even the name 'Webmate' was sort of playing a joke," Marisa explains, on those "crappy" helpful avatars companies were starting to put on their Web sites, all shades of Clippy, the dreaded Microsoft Word paperclip.

Here are a few more things you might have found on Word.com, circa 1996: photo essays of Russian prison tattoos and Manhattan clocks. Bawdy paper dolls. Found thrift store paintings. An interactive game in which the objective is to pop zits. Dean Martin horoscopes and clip art comics. Meaningless interactive toys, like a cartoon microwave you drag a dog into. Depressing children's stories. Photographs of empty billboards and road signs taken across the country. Scanned cocktail napkins with phone numbers and the sad, sexy stories of the people who scrawled them. At least one first-person account from someone who got a coffee enema.

Word was not the first magazine on the Web, but it was the first to demonstrate the creative possibilities of online publishing—and its formula worked. By 1998, *Word* was logging ninety-five thousand daily page views, and one to two million hits a week, which amounted to massive traffic in those days. *Newsweek* announced *Word* as its readers' favorite online destination, and the *New York Times* used the site as an example in one of its earliest articles about Web browsers. Even the first commercial Web browser, Netscape, which was used by essentially every person online in the mid-1990s, gave it a place of privilege. Netscape had a row of buttons beneath the URL field: "What's New?," "What's Cool?," "Destinations," "Net Search," "People," and "Software." "People" led to an early Internet white pages, "Software" to Netscape's own upgrades. "What's Cool?" pointed to Word.com.

Everyone at *Word* shared a certain basic aesthetic: *Twin Peaks,* zines, pop culture, weird home pages. Marisa once compared herself and Jaime to the "Beavis and Butt-Head of the Internet," and the magazine "functioned more like a rock band than a publishing entity," with Marisa and Jaime as frontwomen. "Jaime and I had as much a chance as anyone, just the two of us, at doing something good," says Marisa. Where Jaime was obsessed with multimedia—"She would literally *hiss* about text," Marisa says, and laughs—Marisa knew how to cultivate writers. One particularly dynamite photo of the two of them from this era might as well be a band photo: on the rooftop of the *Word* office, Jaime in combat boots and track pants, Marisa in a leather jacket, a smoking computer at their feet, screen smashed to bits. But as in any good rock band, egos ran wild.

In the escalating fever of the dot-com boom, it was hard for Jaime to stay centered. Silicon Alley was changing fast. Friends who had been techno-bohemian party people only a few years previous were becoming millionaires. Right before joining *Word,* Jaime had turned down a one-third stake in a Web development shop, Razorfish, that was now on the eve of a $1.8 billion IPO. "We knew people who were getting filthy rich," Marisa says. With opportunities passing left and right, Jaime started to get itchy.

And there were other tensions. The public persona Jaime had been cultivating in the press for so long, as a rule-breaking digital zinester, was starting to get on her coworkers' nerves. "All the attention was on me," she admits, "instead of on what it should have been." She was a strong brew, always blasting her music loud and gunning to work on her own stuff. She wandered away from *Word* after a year and a half to pursue freelance contracts on her own. Although a Silicon Alley gossip column at the time pegged Jaime's departure as a power move—"Jaime Levy has the last Word"—Jaime wishes, now, that she could have hung on a little longer. "I've never been a person to last anywhere with anybody," she says.

Marisa stayed on. It hurt to lose her partner in crime, but the idea of leaving *Word* was unimaginable. "It was the fulfillment of every creative dream, and urge, and idea, that I'd ever had," she says. "To pass that up, I just felt would be like cutting my heart out."

THE DEATH OF THE WEB AS WE KNOW IT

At the tail end of 1996, Jaime Levy threw her last CyberSlacker party. She went all out for this one, upgrading from her loft to the Clocktower Gallery in Tribeca and asking some of her more illustrious Alley friends to come onstage and "do rants." As New York avant-garde stalwarts like Laurie Anderson and Lou Reed mingled in the crowd, Nicholas Butterworth, a former punk bassist whose music Web site had been bought by Viacom, wailed for fifteen minutes into the mic. "It's the death of the Web as we knew it," he screamed. "It's over! And wasn't it good while it lasted? Who was there, who was there in 1995? Reaping it in—the money, the fame, the parties, all of it flowing in. . . . It was beautiful! Everyone here, all my friends, doing creative things with a capital C . . . the dream was to be a media assassin, to be a guerrilla—and to be paid! Well, let me tell you something: you now have a choice. You can be a guerrilla, or you can get paid. You cannot do both."

Jaime chose to get paid, but she ended up being a guerrilla.

She left her mark on *Word*. The magazine kept her flash, her "pooper-scooper" interactivity, and her choice hires. Beyond Marisa, there was Yoshi Sodoeka, who filled Jaime's shoes as art director. Yoshi "was one of these Japanese kids that had come to the United States because he wanted to do punk rock, originally." Like basically everyone else at *Word*, he'd never worked in publishing, but he had designed interfaces for kiosks back in Japan. The result was a pop graphic sensibility that earned the magazine its share of copycats and Yoshi his share of lucrative job offers, all of which he declined ("He's an artist," Marisa says admiringly). Yoshi was the first to make icons a core element of site design; even today, his art direction holds up.

Word had an office in Midtown, in the Graybar Building above Grand Central Station. Compared to other Silicon Alley workplaces, it was relatively anonymous. "We were essentially artists and bohemians, not *trying* to be hip," Marisa remembers. She instinctively mistrusted the playground offices of their peers: all the Ping-Pong tables, beanbags, and fancy conference rooms felt suffocating. Of course, *Word* was a product of the boom, too. Marisa doesn't remember if the site even made money, because although its sales department sold ads at around $12,500 a pop in the early days, it largely functioned as a "sexy PR toy" for its parent company.

"For us, the creative team, it was like we were pulling a long scam," one longtime *Word* employee, Naomi Clark, explains. Despite *Word*'s heartfelt authenticity and its experimental design, the site existed "as a sort of trophy wife for these much less glamorous Internet businesses that were in the process of selling infrastructure and T1 lines." At the height of the bubble, all that really mattered was to generate some healthy interest before an IPO. "Going public was the business model, period," Marisa concedes, and Icon CMT wanted to "get the killing and get out" just like everyone else. Like many online publishing concerns with no clear revenue model, *Word* was precarious from the outset.

By the late 1990s, technology investors finally hipped to the fact that although "content is king"—this being the unofficial mantra of Alley types in the early years, so much so that *Word* staff printed ironic

T-shirts reading CONTENT PROVIDER in big block type—the increasing volume of free content on the Web would drive advertising rates into the ground, and nobody was going to pay for subscriptions to digital magazines. *Word* was brusquely shut down by its parent company in March 1998, prompting a flurry of obituaries. A competing online magazine, Suck.com, pondered its "D. B. Cooper–ish disappearance." *Wired* called it the end of an era. But by September, it was back from the dead.

Here's where the story gets weird. Being a trophy wife on the arm of an infrastructure company was one thing, but the holding company that bought *Word* for two billion dollars in 1998 made *industrial-grade fish meal.* "It made so little sense," Naomi says, and laughs. "They had a slightly misbegotten idea that they could start up a media business, and having some sort of known brand names in the mix would be useful for that." Such odd bedfellows are less startling in the woozy context of the bubble. Their new owner, Zapata Corporation, was founded as an oil company in Houston in 1953 by a young George Bush before being retooled into a food processing business by a second owner. In the tech bubble, it was reinventing itself once again as zap.com, a Web portal for the content coast. Zapata's CEO, Avram Glazer, took out full-page ads in the *New York Times*—ZAP WILL BUY YOUR WEBSITE—and entered into negotiations to purchase thirty. *Word* was their marquee acquisition, signaling to Alley investors that Zap was a serious Internet company.

Marisa hired back her staff, all the "underachieving subgeniuses who smoked too much pot and took too much acid in high school." The band was back together. Even Fred the Webmate returned for a second season. In the new installment, he was back at work, too, pacing the halls of a corporate office, working data entry. Visitors could still chat with him, only now the conversations took place while Fred hid in the bathroom, drank Cokes, and did push-ups in the office kitchen. He claimed to be happy and said he'd have a better job soon, but his mood swings were still violent. Maybe Fred would never really get what he wanted.

But on the eve of Y2K, on its last, glorious lap around Silicon Alley, *Word* did.

ELECTRONIC HOLLYWOOD

In 1998, Jaime Levy was invited back to NYU to give a lecture to students about Electronic Hollywood, the company she'd started a few years after leaving *Word.* She tells me what she told the university: "I was like, okay, why don't we call it 'How to spend a half a million dollars in one year and all I have to show for it is this stupid cartoon'?" She went down to campus "totally depressed, probably between my cold-turkey cleanouts," with a mind to tell everyone about her mistakes.

The years after *Word* had been an object lesson in not believing your own hype. Jaime had assumed that when she left the magazine, she'd find another high-paying creative gig, but the landscape had changed. The bankers and investors had really come to town by then, and the cyberpunks were combing their hair and cashing out. Gone were the creative pushers, replaced by people making commercial Web sites, making—and she says this often, relishing it—*Tampax dot com.* At thirty, Jaime wasn't precocious anymore. As she tried to scrap together freelance gigs, her annual salary plunged. She borrowed money from her dad for therapy. She spent a year designing a series of dystopian chat rooms based in a post-apocalyptic version of San Francisco, decorating them with images of burned-out tech companies, radioactive burritos, and zombie drug pushers. In a 2000 *Dateline* profile, she calls it her Kurt Cobain crisis. Then along came that half a million.

She benefited from someone else's stroke of entrepreneurial success. A friend she'd met back in the *Cyber Rag* days at an electronics trade show sold his company to Microsoft; newly flush, he invested enough money in Jaime for her to open her own shop, which she named after one of her later floppy disk magazines. Electronic Hollywood was what she called a "production studio for the Internet," a multimedia agency that produced interactive Web projects, video content, and animations for the highest bidder. It wasn't a huge company, but the office was a digital kibbutz, a *coworking space* long before there was a term for it. When clients came in, they'd see twenty or so weirdos in the conference room: friends who rented desks and threw down programming and design help whenever Jaime needed an extra hand. The office was a fake-out, but it worked.

For Jaime and the early true believers, the fake-out always worked because the objective reality of their industry was impossible to pin down. Clear across the country from the Silicon Valley chip pushers, knowing how to code HTML made you a guru, and there was no shortage of corporate clients willing to pay top dollar to have the Web explained to them. For a 1998 "Silicon Alley Talent Show" at Webster Hall, where a group of the early true believers performed to raise funds for upstart Web projects, Jaime expressed this ongoing hustle in a rap:

Back in the day when new media was new
I could bullshit my employer 'cuz no one had a clue
I was making e-zines on my Mac II
I was totally wired, not like the rest of you . . .
I'm the biggest bitch in Silicon Alley
I'm better than the nerds in Silicon Valley
Bill Gates calls me up when he needs advice
'Cuz I'm Jaime Levy and I'm cold as ice . . .
Now I'm a CEO running the show
I said: now I'm a super HO running the show
Now I'm just waiting for that big IPO

Not long after, she gave her dejected lecture at NYU, having gone from a braggadocious CEO to a depressed startup founder in less than a year. In Silicon Alley, ultimately, even the biggest bitches weren't impervious to the market.

How do you spend half a million dollars in one year, anyway? You start with five hundred thousand dollars in venture capital. You hire your brother. You hire your ex-boyfriend. You sign a five-year lease on an office space with a T1 line—an expensive, high-speed Internet connection. You hire your friends. You rent desks in the office to even *more* friends, cheap, in exchange for help on projects here and there. Every other startup in Silicon Alley is gunning for a big-money IPO or making

Web sites for banks, but you focus on content. You pick up some commercial projects on the side, making interactive toys for companies like Kraft, Tommy Hilfiger, and Bounty. But all your real effort goes into your pet project, the thing you think will pull you out of the tech bubble: a sixteen-episode Flash cartoon series. Because you used to throw a great party by the same name, you call it *CyberSlacker.*

CyberSlacker is a semiautobiographical cartoon about an East Village hacker chick. In three-minute episodes, it tells the continuing adventures of a peroxide-blonde misanthrope named Jaime—"but I go by CyberSlacker online"—as she navigates the surreal New York tech scene. In one episode, CyberSlacker tries to get a programming job. First, she calls IBM, where Jaime worked for a year in 1993, but the milquetoast nerd on the other end of the line gives her the creeps. Next she calls MTV, where a spiky-haired bro answers the phone: "Welcome to MTV's online spankin' new media division-aro, you ready to party?" Finally, she tries a Web development company, Blowfish, a thinly veiled reference to the hotshot Alley startup Razorfish. "I'm just looking for a job somewhere normal," she tells the secretary. "Someplace where everybody's not in some new media Silicon Alley trendy freak-out."

Jaime recorded the voice-over for *CyberSlacker* herself, and the series is so specifically referential to New York's technology community in the late 1990s that it's become a Rosetta Stone for people like me. At Blowfish, stock prices spool along the walls, and even the receptionist evangelizes about all the dollars looking for a home online. Everyone in the office wears matching Steve Jobs mock turtlenecks. The CEO yammers random Internet buzzwords. He thrusts a container of the company's "secret sauce" into her face: it's "Blowfish's own pioneering broadband solution-oriented end-to-end flavor enhancement to spice up the Internet," he says. But CyberSlacker sees that the emperor is naked. Her retort: "It looks like a jar of salsa on top of an old PalmPilot." She smashes the secret sauce and goes on an anti-capitalist tirade.

For the early true believers, the Alley obsession with making money was beginning to feel offensive on a spiritual level. After Jaime's own hot flash of entrepreneurial fever, she learned not to trust anyone who chose a quick buck over interesting work. She turned down a one-third stake in Razorfish in 1993; it went public in 1999 and was worth $1.8 billion, employing nearly two thousand people in fifteen offices around the world and wanting so little for work that it sometimes fired *clients* for not being cool enough to be represented by the agency. Their slogan was "Everything that can be digital will be."

Things did not go so well at Electronic Hollywood. The overhead was too expensive, and Jaime quickly realized that she should never have hired her friends. She eventually had to fire her own brother, who'd moved from the West Coast to work for her. Her ex called Electronic Hollywood's investor to tell him personally that Jaime was smoking pot in the office. "Everyone else either quit or turned against me," she said. "We pissed away almost all the money." But Jaime held on to the *CyberSlacker* cartoon as her potential ticket out of the Alley's trendy freak-out. She wanted to have something to show for all that money. She was certain she could take *CyberSlacker* to television. In the echo chamber of the Alley, Jaime plugged her ears, and looked for the exit.

SMALLEST VIOLIN FOR THE FUCKING CYBERKIDS

Jaime never did get that big IPO, and the *CyberSlacker* cartoon didn't make it to television. To her great misfortune, Jaime founded her company on the eve of the stock market crash. Not long after rapping about being the "biggest bitch in Silicon Alley," Jaime felt the air change. "Something was coming to a head," she remembers. "It had already been too many years of too much money for no business model, no return on investment." In late March 2001, the financial newspaper *Barron's* ran a story listing two hundred Internet companies running out of money, and within a month, the market was losing between one and three hundred points a day. On March 12, 2001, the NASDAQ dropped below 2,000 points, almost a year to the day from its all-time peak of 5,048.

The cover of the *Silicon Alley Reporter* that month said it all: a black-

and-white photo of the Hindenburg, consumed by flames. The party was definitively over. "Within two months," says Jaime, "everybody fired everybody." Pseudo, the live-streaming service that had bankrolled Josh Harris's lavish parties, shut down, leaving 175 employees to wander dazed into the light of day; a former art director, quoted in the *New York Times,* said, "We ate from the trough of the venture capitalist pigs . . . now I'm crawling back to the corporate dog bowl." Razorfish, the Web design agency Jaime had satirized in her *CyberSlacker* cartoon, didn't fare any better: as the company's stock teetered on the brink of delisting, Razorfish ousted its founders, and a sociologist visiting its Soho offices in 2001 compared it to "one of those tombs built for Chinese emperors," with hundreds of empty office chairs like "rows of terracotta soldiers."

Electronic Hollywood cleared out its soldiers, too, leaving only Jaime and her office manager, Maria. Their main equity was the office space itself: with six months left on the lease, Jaime needed the $20,000 deposit back if she was going to have any chance of surviving the crash. She booted the cyberpunk squatters and moved Electronic Hollywood to a small room in the back, renting out the rest of the building. "This is going to be our game plan for the next six months," they told each other. "Just survival. Keep this small space, which is all we need." But there was no work, Jaime was burned out, and Electronic Hollywood was down to one client. She and Maria committed to finishing off the lease and moving on with dignity.

The money had never been real, and now it really wasn't. The crash turned multimillion-dollar stock holdings into worthless paper; the big Web design firms, Razorfish and Agency, lost 90 percent of their value as it became clear to investors that having a Web site was not equivalent to having a business model. Every company doing business or making content for the Web was fighting for its life.

Marisa Bowe saw the crash coming a mile away. Being from the Midwest, she'd lived through the collapse of the steel industry. For her, being part of the breathless atmosphere of the bubble was a rush precisely because she knew it would pop—it was like being on the inside of history. Her two favorite novels, Guy de Maupassant's *Bel Ami* and Honoré de Balzac's *Lost Illusions,* take place in Gilded Age France, and to be living through such a folly herself felt like a crazy stroke of luck. "Normally someone like me would be on the outside just reading about it in the newspaper," she says, "but through a weird combination of circumstances I actually *knew* people who were getting rich, and just to be on the inside of it was like living in one of those novels that I loved. It was just incredible."

Jaime Levy didn't take it so well. The crash represents a hard line in her life of hustle, bombast, and creative entrepreneurship. When the Alley fell, so did the community that had forged and made credible her reputation. The print media outlets that had hastened the rise of the dot-com kids with breathless profiles vaulting them to celebrity status caught none of them on their way back down. The fallout was brutal. The big talk, the parties, and the flash that had defined the excitement of the boom suddenly looked like so much excess—and the early true believers, with their fast money and promises of a boundless information future, like charlatans.

And then, like a kick in the guts: 9/11.

People think 9/11 was one day. "That's bullshit," Jaime says. "It was one entire *year,* every single day, walking outside your apartment in the East Village and there's like, a family posting signs: 'Have you seen my mom?'" There were "obituaries every day, people walking around in gas masks." The reality-altering scale of the attack didn't just make Jaime's problems seem irrelevant—it made them seem petty, even cruel. "Smallest violin for the fucking cyberkids who all of a sudden aren't getting paid money to fucking have sushi parties," she says. "Three thousand people lost their lives, including firemen, and the city's covered in smoke." There could be no greater reality check.

As one historian put it, the collapse of the city's great monuments, "coming after the collapse of its grandest illusions, had the epic quality of myth." It crushed the soul of New York and poisoned those who rose to help the helpless. It damaged nearly thirty million square feet of office space and cost 138,000 jobs, sending the city's economy into a tailspin—by 2003, unemployment in New York would hit a staggering 9.3 percent. For many, it was the unequivocal end of the road. At thirty-four, Jaime said good-bye to Electronic Hollywood and moved back to Los Angeles. She got an apartment in Silverlake and a car. She started over. The stink of the crash was on her, and when she started applying for jobs again, she omitted from her résumé her tenure as the CEO of a dot-com. Instead, she told interviewers that she'd been a front-end designer at her own company.

THE SOFT ECHO OF THE BOOM

When Zapata, the fish-meal processing company, finally shut down *Word,* it did so very quickly. Nobody had time to save anything. Jaime had stolen a backup the last time she was in the building, but she got "fucked up and left it on the subway." Marisa took a more enlightened approach. "I started thinking of it as being like those monks who make the sand mandalas," she says. "They work for days and weeks to make this perfect thing and then the wind just blows it away."

A Web site ought never hope for permanence. Left untended, all bits eventually evaporate. On the World Wide Web, links rot—a consequence of its one-way hypertext design—images disappear, and any desirable URL not held on to tenaciously stands little chance of survival. Which is why if you go to word.com today, you'll find a dictionary and not a magazine. The real estate is too valuable to host memorials.

The early Web publications were replaced by more magazines, and then personal blogs, and eventually the social media platforms that hybridized magazine and community by selling community to the advertisers. As they disappeared, with them went hard-won triumphs in the new medium, what Alley businesspeople already called "content" back in the 1990s. As the editors of Suck.com, one of *Word*'s rival Web publications, wrote as they contemplated the disappearance of yet another online magazine in 2001, "We figured that in the end there'd be something left after all this effort . . . something, anything, rather than just this Krakatoa memory, echoing across an ocean whose smallness we're just beginning to fathom."

The crash didn't just gut an industry. It took down the generation of creative people who found miraculous employment doing what they loved, and who, in the process, defined the cultural parameters and interactive possibilities of the Web. The money survived—in some form, it always does. But the artifacts of the culture it briefly enabled are harder to find, especially without a map.

You might wonder why anyone should care about the Krakatoa memories. *Word* has been off-line far longer than it was ever online. Jaime Levy's floppies are sitting in her home office, although she's working on getting them restored and hopes to show them in a museum someday. Echo, an anachronism in the age of big social, has become the Internet's smallest local bar. These places, built by bright-eyed true believers, by women who were there right from the beginning and who didn't make a dime for their labors, are just as important as the fortunes made and lost in the feverish speculation that transformed the Web from an academic swap session to the beating economic and cultural heart of our world.

We should care about early online communities and publications, as we should care for their archives, because they were the places where the medium revealed itself. The financial bets being made on the sidelines are remembered because of their dark reverberation on the economy. But another bell tolled, and it still rings out, growing fainter by the day: the contributions of those who saw the Web's potential right away. After all, the Internet's only job is to shuttle packets of information from one place to the next without privileging one over the other. Our only job is to make the best packets we can. To make them worthy of the technology.

Marisa Bowe made great packets, alongside her friend Jaime Levy. They played the Web for what it was: interactive, homegrown, human, funny. Stacy Horn took no greater pleasure than in filling her servers with voices. She helped people, especially women, get online. She and her Echo hosts taught themselves to manage the crowd and make sure everyone was heard, and they grappled with questions of gender, privacy, and responsibility online that we still haven't quite answered. None of them got rich, and much of what they built has been erased in the slow erosions of the Web, like *Word,* is walled away from the Web, like Echo, or else is bound to media inaccessible to anyone but digital archivists with the means to replay them, like Jaime's floppy disk magazines.

This makes their achievements difficult to remember, much as the programs patched by women's hands in the basement of the war theater a generation previous are difficult to remember, or much as the hypertext systems that could have become as important as the Web had they only been implemented on a larger scale are difficult to remember. These all exist in fluid configurations of time, relationships, and clicks. These all exist in relation to the ugliest things in our world: the trajectories of bombs, the frantic pursuit of wealth. But all of their efforts are artifacts, as the Web is a dynamic artifact, an endless conversation writing over itself again and again in the soft echo of the boom.

Chapter Twelve

WOMEN.COM

The dot-com bubble began to inflate with the earliest home pages and Web magazines, and it didn't pop until the Web went fully mainstream, right before 9/11. In those eight years, a technology that began as a networked hypertext system for particle physicists became the world's gossip page, multimedia art gallery, and library, in a feverish burst of cultural activity the likes of which the world had never seen. No longer was the Internet the realm of computer scientists, academics, undergrads, and the occasional overworked librarian. It was a popular medium, as transformative as television and far more intimate, connecting disparate strangers one-to-one.

But while the first generation of artists, coders, writers, and navel-gazers made meaning from the global communications network, something else happened, too: the Web became a commercial medium. Once Web enterprises figured out their business models and how to securely process credit cards, clicks turned to dollars, forging some of the most powerful companies in the world, companies that have since become titans in distribution, media, and even space travel.

Commercialization changed the Web for everyone. It had an indelible effect on the kinds of sites being built, and on the nature of the content being distributed there: less "my name is Lisa and here's my dog" and more Pets.com. Where Echo managed to remain community focused by charging for its service, one of the few ways that social platforms and content sites make money on the Web—and today, in social apps—is by turning their users into the product, selling demographic information and targeted ad space to advertisers. Bubble Web sites like *Word,* kept afloat in the swell of new money, were forced to fold without a long-term business plan. But other Web sites, more eager to apply traditional media advertising models to the Web, thrived for a time.

Few of these tell a clearer story than women.com. It's one that spans a decade, streaking across two coasts and the Internet's most dramatically transitional years, the story of a grassroots feminist community, the first online destination explicitly for women, which grew into one of the most spectacularly successful media companies of the bubble years. In the process, it became a symbol: first a stock ticker, then a cautionary tale, and finally an avatar for the very soul of the Web.

WOMEN'S WIRE

It starts back on the West Coast, with Nancy Rhine, the former communard who left her hippie roots to homestead online at The WELL. By the early 1990s, her frustration with the BBS boys' club had become a fever. Posting in The WELL's women-only conference, she began talking about "wanting to get an online community started that was focused on stuff that was particularly of interest to women: women's health, or FDA loans for women-owned businesses, or mothering, or anything. Because we're interested in everything."

In that conference, she met Ellen Pack, a New York transplant with a business degree and a nest egg. Ellen had moved to Palo Alto with a startup and settled in Silicon Valley. Nancy lived out in Marin, the opposing pole of the Bay Area's techno-social world, but they found common ground in the idea that women deserved a corner of the Internet all their own.

At the time, there were female-centric spaces online, but they were fairly proscribed, like Echo's invite-only WIT conference and the similar conference on The WELL where Nancy and Ellen met. Women made up only a very small percentage of Internet users. Ellen and Nancy were willing to bet they were just waiting for the right opportunity to come online, the right number to dial. They sketched it out together, and Ellen bought the servers. Their network, the Women's Information Resource Exchange, WIRE—renamed Women's WIRE after a legal dustup with *Wired* magazine—launched to a founding group of five hundred members in October 1993, becoming the first commercial online service explicitly targeted to women.

They designed it to be as approachable as possible. Women's WIRE ran on relatively user-friendly BBS software called First Class, with a graphical point-and-click interface and plain English typed commands. New users were courted by direct mail: the Women's WIRE brochure, covered with images of Renaissance naiads, exhorted women to sign up and "access information and resources instantly, discover new heroes, tell your stories, vent your frustrations, get advice, in short . . . get connected." If they sent away for a starter kit, they'd receive a floppy disk and some noncondescending information about how to set up their modem, and they could always call Women's WIRE customer support—all women, naturally—if the directory names, baud rates, and disk densities proved inscrutable.

Nancy and Ellen received immediate media attention for their effort. "The day we launched we got a front-page, above the fold, article in the *Merc*"—the *San Jose Mercury News,* Silicon Valley's daily paper—Ellen remembers, and "it was like: 'Women Build Site to Keep Out Cyberbores.'" The headline was actually more unbelievable: WOMEN AIM TO BUILD AN ON-LINE WORLD THAT EXCLUDES BOORS, CYBERMASHERS. Ellen was horrified by it. "I remember being devastated," she says. "I was like, '*Oh my God,* they totally misunderstood why we built this.'" Nancy and Ellen didn't start their women's network as a safe space for a fragile population of online newbies. It wasn't a fortress designed to keep out bullies—or God forbid, *cybermashers*. It wasn't even anti-male: men would eventually make up 10 percent of Women's WIRE users. Rather, "it was much more of a positive place," says Ellen. "It was not like we needed a protected island, it was like, holy crap, there's so much we could do with this. Let's do this."

Women's WIRE users quickly realized the power of immediate access to a community of women. The service hosted domestic abuse resources, contact information for female members of Congress, listings for educational financial aid, professional organizations, and women-owned businesses, and transcripts of First Lady Hillary Clinton's speeches. In public forums, users shared advice with one another—not the "how-*to*" of rarefied experts and magazines, but the more personal "how *I* . . ." that we now see on support forums everywhere online. In 1994, when a draft of the Clinton administration's health plan was revealed not to include obstetricians and gynecologists as primary caregivers, two Women's WIRE users sounded the alarm and organized a "phone assault on the White House." Three days later, they posted a victory notice—it had worked. Some twenty years before social media clicktivism, Women's WIRE subscribers mobilized an online community to take real-world action for change.

Women's WIRE had an office, a carpeted second-floor space on a working-class street in South San Francisco, halfway between Marin and Palo Alto. The staff was all female: there was Phyllis, a lapsed Mormon who ran IT; Roz, who'd later become a Buddhist priest; and Susan, with a PhD in linguistics. Nancy hosted face-to-face user meetups in the common space, and when the media came knocking, Ellen and Nancy would pose for photographs in the tawny hills above South San Francisco, holding each other close as the restless winds blew their hair and the tall grasses sideways toward the sea.

But despite their shared goals, Nancy and Ellen were as different as two founders could be. Nancy, two decades older than Ellen, had deep roots in community building, online and off-, and came from a world of hippie self-reliance and the consciousness-raising culture of second-wave feminism. As a reflection of those values, the original Women's WIRE system index included information about abortion, women's studies programs, and lesbian parenting, as well as a direct link to groups like the Boston Women's Health Book Collective and the National Organization for Women (NOW). The way she saw it, Women's WIRE was a radical network, a space for and by a feminist community, which incidentally included its share of men. She imagined a network of voices, anchored by a group of founding subscribers who would "foster diversity . . . people with something to contribute, personally, intellectually, or through their involvements."

"I liked the community piece, when you're sharing useful information," Ellen remembers, "but I wasn't as much of a community person, for community's sake." She was of a new generation of entrepreneurial feminists, with her eyes on the boardroom. As a user, she was more interested in straight information; she wanted to get answers from newsgroups and databases, rather than through the more imprecise channels of anecdotal sharing. As a result, Women's WIRE had two distinct poles, both an *Information Resource* and an *Exchange.* The balance between these two priorities was uneasy. Women's WIRE's publicist, Naomi Pearce, remembers an ongoing disagreement about the nature of the medium. "We had this big raging debate about: Is the Internet about information or is it about community? And I'm looking at this whole argument, going, duh. *Both.* At a certain level of intensity in an either/or argument, the fact that it has reached that intensity is the indicator that the right answer is *and.*"

For fifteen dollars a month and an hourly connection fee, Women's WIRE subscribers enjoyed access to both: there were daily headlines and newsgroups, but users also accessed e-mail, resources hosted by women's organizations, and sections dedicated to "herstory," finance, technology, parenting, and education, among other topics of interest to women. Hangout, the general forum welcoming newbies into Women's WIRE's ongoing conversation, became the service's traffic magnet. "I used to feel like women at home raising their kids would get really isolated," Nancy says. "Where is the quilting bee? Where's the agora? Where's the town square where we're all hanging out?" As she monitored the chatter, she congratulated herself on her instinct: just as she'd expected and hoped, the quilting bee had moved on-screen.

Many early digiterati believed that online conversation would flatten differences between people. After all, few technologies have cleaved words from flesh so thoroughly as the Internet. Perhaps computer conferencing would create a "civilization of the Mind," one proponent went so far as to proclaim, a place where race, gender, ability, and class would finally become immaterial. The reality was quite different. As Stacy Horn learned in her years running Echo, people brought everything with them into cyberspace: their social conditioning, their hard-won truths, and all their baggage.

Women's WIRE, as a rare example of female-dominated cyberspace, was proof. Where The WELL had been a free-for-all for jocular intellectuals, a Wild West rewarding conversationalists willing to own their own words and stick to their guns, dialing Women's WIRE felt like visiting another country. With its inverse user demographics—90 percent female, 10 percent male—the tone of conversation was widely deferential and supportive. The WELL "had a lot of alpha males posturing," Nancy says. But the women on her service "were so polite and nice to each other that sometimes we had to stir it up in order for it to be an exciting conversation."

WORKING ON A ROCKET SHIP

Nancy Rhine and Ellen Pack were running Women's WIRE out of their South San Francisco office the first time they saw the Web. It was 1994, the year the first Web browser, Mosaic, ceded to the successor, Netscape, which would soon become the standard for early true believers on both coasts. Like AOL, CompuServe, Prodigy, and Delphi, all subscription services cut off from the hypertext Web, Women's WIRE was at a crossroads: they could hope that their existing service could sustain itself, or they could redesign everything for the Web.

This wasn't the obvious choice it might seem to be today. The Web was clearly important, but it was risky. Staking a claim there would mean abandoning the thousands of paying subscribers they already had. And in the dial-up days, Women's WIRE was beholden to nobody save its subscribers. Nancy explains to me that it had been possible, before the Web, to run a successful business and a successful community at the same time. In fact, one reinforced the other. "The original model had been how long people *stay* online," she tells me. "You wanted community because that kept people engaged." It's worth noting that modern-day social media giants have come around to this perspective, albeit on a much larger scale; when Facebook introduced private groups in 2010, it was to capitalize on the deeper ties forged in interest-specific online communities, which have become engines of user engagement for the platform. As *New York* magazine put it in 2017, "Facebook is good because it creates community; community is good because it enables Facebook."

Back in the early Women's WIRE days, however, Ellen was less inclined to see the inherent value of online community—sustainable as it may have been, it was not lucrative. She'd come to Silicon Valley with an MBA, and saw Women's WIRE as a startup like any other. She was keen to build something larger than a carpeted second-floor walk-up in South San Francisco could contain. In 1994, the year the Web broke, Ellen started rooting around the valley for venture capital to lay the groundwork for a Web play. "I always wanted to make this into something big," she tells me.

While she was out pitching, she was introduced to Marleen McDaniel, a thirty-year industry veteran, one of the few women at the top of the Silicon Valley food chain. Marleen was a technology industry Zelig: she cut her teeth as a computer girl, a systems analyst and programmer, in the 1960s and spent the 1970s hanging around with the Xerox PARC beanbag crowd. She worked, for a time, with Douglas Engelbart, Jake Feinler's Stanford mentor, on a failed attempt to commercialize his system. In the go-go years of the 1980s, she joined Sun Microsystems as employee number forty and rode it all the way to the top, "like working on a rocket ship," she says. By the time she met Ellen, she'd participated in the successful launches of several high-profile startups and was well respected in the valley. Ellen sent her a business plan and a bouquet of flowers.

At the time, Women's WIRE didn't have quite enough subscribers to be solvent, and Ellen had already sunk her life savings into the company. She and Nancy were having a hard time getting through to the VC firms they were meeting with, because most potential investors were men who regarded the notion of a women's network as little more than a curiosity. "Some got it, of course," remembers Ellen. But with others, "you'd get the classic, where someone says, 'I'm gonna have my wife look at this.' They didn't really take it seriously as a market or as an opportunity. They would think we were a fringe group." With Marleen in the picture, however, there was a chance Women's WIRE would be taken seriously long enough to make their case.

"It was not easy to raise money for this company," says Marleen. In her estimation, it wasn't a gender issue: the Women's WIRE business plan wasn't forward thinking, and First Class BBS was on its way out. But the brand had first-comer status, and the idea of using the Web as a commercial medium was still tantalizingly new, so Marleen came on as a consultant. She went out pitching with Ellen and Nancy, but her credibility had an unintended effect. "When I finally got a smaller but high-quality VC firm to make a commitment to it, they did it with the caveat that they wanted *me* to be CEO," Marleen says. "So that was the first crisis."

Marleen is an inveterate early adopter, the result of a career spanning several generations of computer development. She knew that Women's WIRE couldn't survive as both a content destination and a service provider, competing with giants like AOL, which had the resources to mail millions of installation disks to potential new users around the country. Having first encountered the Web in the early 1990s, she knew it was the only play for Ellen and Nancy, despite the risk. "That was a defining moment," she says, "and in case you think that starting a company is easy—it's not. This was like jumping off a cliff and having no idea if there's water." Only one thing was certain: Women's WIRE was no longer a community service. It was a media company.

Soon after, Women's WIRE announced that it would abandon its stand-alone dial-in service to focus on creating "electronic magazines" for Microsoft, CompuServe, and the World Wide Web. The first big deal

Marleen brokered was a multiyear arrangement to divest the Women's WIRE community from their platform by urging users to switch to CompuServe's $9.95 monthly plan. "This is an opportunity to build a large, diverse, and culturally rich women's online community," she wrote in a system-wide e-mail announcement. ("They bought our subscribers," she tells me now.) As these changes rippled through what remained of the original Women's WIRE community, Nancy walked away from the company, taking an exodus of users with her. "They're saying it's a partnership between CompuServe and Women's Wire," said one forum moderator of the forced migration, "but it's clear to me that the Women's Wire that I signed up for hasn't existed for several months. This is a shifting away from a community model to that of an information provider." On Halloween 1995, the old Women's WIRE modem lines went down.

Nancy kept mum at the time, but she can't have been pleased. She came from a generation of early adopters who believed that the Internet was inherently a community technology—to her, subscribers were citizens before they were consumers. Nancy's old-school, egalitarian vision made no sense to Marleen. "She wanted to run the company," Marleen says. "Instead of having a CEO and a pyramidal structure . . . she wanted to have sort of a Knights of the Round Table kind of situation."

Marleen spent 1995 securing distribution partnerships with companies like MSN, Bloomberg, and Yahoo!, and she and Ellen launched Women's WIRE on the Web that August. The company's central online service became a content destination—more magazine than sewing circle, a place where women could *get things done,* from checking their stocks to reading their horoscopes. Articles written by Women's WIRE staffers and freelancers were laid out among stories pulled from the Reuters newswire, alongside forums on subjects "ranging from Barbie to Bosnia," as Ellen told one journalist shortly after their launch. For all intents and purposes, Women's WIRE became a magazine. "The model switched to eyeballs and ads," says Nancy, and "the emphasis really became just running people through. It didn't matter how long they stayed, and whether they were engaged or not. It was just selling eyeballs like a magazine."

GETTING THERE

With their first infusion of venture capital, Women's WIRE moved from South San Francisco to a nicer office in San Mateo. Ellen was deputized to vice president of Product Development, and she spent her time flying back and forth between San Francisco and New York to meet with media partners and hire the team of editors that would run the new, magazine-like Women's WIRE on the Web. She hired Laurie Kretchmar, a former editor at *Working Woman* magazine, as editor in chief, and poached Gina Garrubbo, a savvy sales manager, from the Discovery Channel. Gina was converted from the moment she first read Women's WIRE articles on the Web. "I said, 'Oh my God,'" she remembers. "They assume women are intelligent. Have money. Are doing their own investment. Are making their own fashion decisions and not wanting to be told what to wear and not to wear. There is no condescension in this voice. I felt that I had never heard this voice in the universe."

Not everybody was so quick to see the potential in building a brand for women online. Gina tells me a story about a 1996 investment meeting. "Marleen, Ellen, me, and our CFO go into a bank in Boston, who will not be named," she says. "We were all dressed in black jackets and black slacks. We all have laptops with us. We walk into this 1980s conference room." Around the table sat seven bankers: suits, white shirts, power ties. "We start telling them how women are currently not the majority of online users," Gina remembers, "but *will* be the majority, how women are going to buy things online, are starting to research things online, and they look at us like we're mad. This one guy was like, 'My wife doesn't even like computers.' They were thinking, *These guys are nuts.* And we're looking at them like, *These guys don't get this at all.*" After the meeting, the women walked out onto the street and laughed. It was all they could do.

The bankers were wrong, and the women knew it. As the 1990s progressed, the Web began to reflect the gender demographics of the real world. Where text-only BBS clubhouses, Multi-User Domains, and newsgroups had been dominated by hobbyists, techies, early adopters, and teenage boys, the Web opened the floodgates for women. By 1998, 39.6 million women were online, and in 2000, the number of women online surpassed that of men for the first time. This shift represented a massive deluge of female users of all ages, leading one prominent market research firm to announce that "it's a woman's World Wide Web." Riding the wave, Women's WIRE changed its name to women.com.

"When we jumped off the cliff," Marleen says, "we had to invent a business model." That business model was advertising. Women.com was first in the market by virtue of being the first on the market. As one of the only female destinations on the Web, its frank, tech-savvy voice promised a refreshing alternative to grocery store women's glossies—stuffed then, as they are now, with sex tips and fad diets. Alongside sites like Aliza Sherman's Cybergrrl.com, a dishy, Gen-X guide to cyberspace, and iVillage, a parenting community turned women's network, women.com represented a new kind of women's media: sincere, connected, and smart. And for sale.

Because women controlled more than 80 percent of consumer purchasing, advertisers paid close attention to the growth of the so-called woman's World Wide Web, and women.com was quick to capitalize. Gina's team sold the first ad on the Web, a section of the site underwritten by Levi's. It was what we now call sponsored content: the Levi's section, "Getting There," profiled women with cool careers, like Webmasters and private eyes, much as *Word*'s beloved "Work" section had explored the extremes of the American workplace. In those early days, large brands were eager to advertise online, but there were no standards for pricing, sizing, and terms, and nobody knew how page views translated to advertising impact—but traffic could be measured, and the traffic was good. While the original Women's WIRE dial-in service had only fifteen hundred paying subscribers, by 1996, women.com was getting 7.5 million hits a month. "We were on the map," remembers Ellen. "Nobody else had female eyeballs like we had."

Like word.com, women.com got an early traffic boost from being featured as one of the Netscape browser's "What's Cool" destinations, but its easy-to-remember URL played a role as well. In the years before Google, Web browsing began with a URL typed directly into the address bar. From there, you'd follow hyperlinks on directory sites like Yahoo! or else explore dedicated "link pages" personally curated by a site's Webmaster (or -mistress). This made a URL like women.com especially valuable: odds were, people would type it into the address bar just to see where it led.

In 1997, women.com expanded to become a network, with a constellation of interconnected sister sites: Women's Wire, MoneyMode, StorkSite, Prevention's Healthy Ideas, Beatrice's Web Guide, and Crayola FamilyPlay. Three of these were underwritten by sponsors, who paid to partner with women.com and cleave off a slice of its traffic, and after Levi's, women.com sold more advertising space to Fidelity Investments, Honda, Toyota, AT&T, Coca-Cola, Procter & Gamble, and American Express. According to an industry research firm, it ranked among the most profitable sites on the Web. The Levi's ad had gone for ten thousand dollars in 1996. Two years later, banner ads and sponsored content were routinely being sold in multiples of a hundred thousand dollars, and, by the end of 1998, in excess of one million dollars a pop. "We built the revenue from nothing to tens of millions of dollars," says Gina Garrubbo, women.com's VP of Sales.

All of this was a long way from Women's WIRE. If the original business had had spiritual grandparents, they might have lived at Resource One: Women's WIRE had Community Memory's organic locality and the Social Services Referral Directory's social consciousness. Where content on Women's WIRE had "emphasized current news and affairs and encouraged political activism," women.com was a different animal, and the evolution from one to the other reveals how quickly early efforts to claim online space for women turned into building businesses that sold products to women, and then to the commodification of the women's Web as a whole. It wasn't long before women.com, a venture-funded company aggressively seeking advertising money to stay afloat, became indistinguishable from the grocery store women's magazines to which it had been founded as an alternative. Despite the leap from pulp to pixel, women.com was no more nutritious than *Cosmo*. In fact, it was *Cosmo*.

TWO CARS ON A RACETRACK

When Marleen and Ellen launched their Web play, they were the only game in town. But the women's Web quickly grew crowded: in 1998, Geraldine Laybourne, a television executive, announced Oxygen Media, a cable channel with an accompanying Web presence staffed by whip-smart Gen-Xers hired away from indie Web zines. ChickClick, which began as a small network of "grrrlish zines," became an affiliate network of fifty-eight youth-leaning women's sites with names like Technodyke and Disgruntled Housewife. But only one competitor kept them up at night: iVillage.

iVillage's CEO, Candice Carpenter, was a tough customer in a power suit. "She was scandalously interesting," Marleen tells me, with a mixture of admiration and something else. Carpenter had grown her venture from an AOL community of kibitzing parents into a thriving knot of seventeen women's Web sites, underwritten by huge venture capital rounds, corporate alliances, big-money advertising plays, and early ventures into e-commerce. iVillage was built on a community model: its marquee product was forums, where women shared everything from postpartum anxiety and breast cancer stories to advice for managing work stress and unruly teenage children.

Carpenter knew that community would keep visitors coming back to iVillage and that it boosted the site's page views, a critical metric in a business with no other discernible assets. iVillage was expert at capitalizing on the repeat traffic its forums encouraged: join a pregnancy support group on iVillage and you'd eventually find yourself on Amazon, browsing baby books, or on one of iVillage's sister e-commerce sites, like iBaby, picking out rompers. Carpenter, whose résumé included stints at Time-Life Video, QVC, and AOL, spoke of "monetizing" the iVillage community, turning the network's social dynamics into a business worthy of the stupendous amounts of venture capital it was known for burning through.

By 1999, iVillage and women.com were "like two cars on a racetrack," Marleen explains, competing for advertising dollars and prominence on the Media Metrix top fifty, the early Web's version of Nielsen rankings. "We competed for everything," says Gina Garrubbo, and in their efforts to outpace each other, iVillage and women.com inched closer and closer to the middle. iVillage, which began as a community site, started publishing articles to lure the content-curious; women.com, seeing the popularity of iVillage's forums, added more of its own. Laurie Kretchmar, women.com's longtime editor in chief, puts it more simply. "There's often two in a category, like Burger King and McDonald's, Pepsi and Coke," she says. "Personally, I like Coke."

In 1998, Marleen sold a large stake of women.com to Hearst Communications—one vote shy of controlling interest—in exchange for a comprehensive brand partnership. The Women.com Networks, Inc. became host to Hearst's female magazine brands: *Good Housekeeping, Town & Country, Marie Claire, Redbook,* and yes, *Cosmopolitan.* This gave women.com an aura of legitimacy, but made its content indistinguishable from the mass-market media the Web was supposed to be disrupting. By the end of the decade, as the combined result of the Hearst merger and the Web's explosion as a commercial medium, women.com had gone mainstream, devoting an increasing share of space to its flimsier magazine partners. The puff had sneaked in, irrevocably: "71 Weight Loss Tips That Really Work!" "Mother-in-Law Driving You Crazy?"

The same thing was happening on women's sites across the Web. From iVillage to Oxygen, feminism ceded to fluff and political commentary gave way to fad diets, horoscopes, and compatibility calculators. The Web was no longer a reprieve from the pabulum of *People* magazine; sex and horoscopes drew more traffic than anything else, and "those who thought the Web would be more like *Ms.* than *Mademoiselle*—believing that all women were itching for more intellectualism—were probably deluded," noted one critic, in *Salon*.

Early efforts to identify and develop the female Web had an inverse intended effect. As the Canadian scholar and theorist Leslie Regan Shade points out, the more aggressively sites such as women.com and iVillage catered to female readers, the more they pigeonholed them, until all that remained of their communities were consumer demographics and ad clicks. This continues to be an issue today, as online media struggles to balance the conflicting needs of advertisers, subscribers, and readers. In the heady atmosphere of the Web's early days, however, it seemed particularly grim: women had come online for empowerment and community, and within only a few years they were being sold jeans, baby clothes, and lotion.

A blistering *New York Times* editorial sounded the alarm in the year 2000: the commercial success of women's online media was symptomatic of a "virulent cultural separatism," a corny, insidious "woman's culture," spreading everywhere from chick flicks to reality TV to fashion mags, designed solely to trick female consumers into loosening their purse strings. "There's the ultimate deception," the *Times*'s Francine Prose wrote, "the marketing-research-driven con, the appalling bait-and-switch practiced on the woman who is being promised relationship, being sold on community, and who is in fact buying into a progressively deeper isolation and seclusion."

FUCKED COMPANY

One of the most interesting artifacts of the dot-com bubble is a Web site, FuckedCompany.com, that translated Silicon Alley's meltdown into a game of fantasy sports. Every day, it posted new industry "fucks," rating each by severity: layoffs, distressing press releases, and empty offices. One hundred severity points and the company was retired to the Fucked Company Hall of Fame, and likely went bankrupt in real life, but bets could be placed on each progressive lurch toward the end. Nervous tech workers used the site as a barometer, and some even played the game themselves. After the turn of the millennium, the only way to make money on the dot-coms was to gamble on the specifics of their demise.

In December 2000, Fucked Company posted a new notice:

> Women.com, my favorite site for reading about what chicks like, laid off 85 beautiful, bikini-clad ladies.
>
> Severity: 45

The women's Web had been riding high in the late '90s, selling million-dollar ads and charting unprecedented traffic numbers. Even as cultural observers were lamenting the sites' descent into puff clickbait, speculative financial interest in the women's Web was at an all-time high. Like many big Internet businesses of the era, both women.com and iVillage felt the pressure to take that interest to the bank with well-timed initial public offerings.

iVillage went first, filing its S-1 with the Securities and Exchange Commission in 1999. Expectations were high. Candice Carpenter was an ace hustler, with an unparalleled ability to portray a company "with deep losses, few physical assets, little proprietary technology, extravagant rates of spending, a high employee 'burn rate,' and powerful emerging competitors" as worthy of substantial investment. On the eve of iVillage's IPO, which opened after an all-time spike in the Dow Jones Industrial Average, she'd already led the company through numerous funding rounds, culminating in a flashy mezzanine round of $31.5 million.

iVillage set its offering price at $25 a share, and when the markets opened, bidding began at $95.88, making Carpenter, her business partner Nancy Evans, and their investors multimillionaires in a matter of seconds. Back at the iVillage office, employees were so delirious with joy that Carpenter told a reporter she was "peeling them off the ceiling." Her own stake was worth eighty million dollars. It would be one of the most successful IPOs in the history of Silicon Alley.

Women.com was watching closely. "I was upset that iVillage went public first," Marleen says. She and Ellen had been tracking their rivals, eager to see how a woman-run Web business would fare in the indiscriminating crosshairs of Wall Street. iVillage's stratospheric success was a good sign, in some ways: it showed that the market took women's media seriously and was willing to stake big money. But Marleen wasn't certain lightning would strike twice. "It affected me," she admits. "I said, 'Oh shoot, we better get out there.'" Women.com Networks, Inc. filed their own S-1 only two months later. "You know when the time is right because the investment banks are coming suddenly to visit," Marleen explains. "It was the right time. It was probably the only window we could have used." That October, Marleen and Ellen watched their new stock ticker spool across the screens in Times Square: WOMN.

Women.com's IPO was nowhere near as splashy as iVillage's, but it was perfectly reasonable: the company's initial $10 offering price per share nearly doubled on the first day of trading, and climbed to $23 by the end of the month. On the back of a high-profile advertising campaign that followed, women.com even nosed past iVillage in the Media Metrix top fifty, prompting iVillage to take out a full-page response in the *Wall Street Journal* implying that women.com had earned its place only by merging with Hearst. THERE'S A DIFFERENCE BETWEEN AGGREGATING WEB SITES AND BUILDING THE LEADING BRAND, read the headline. But despite such antagonism from its primary competitor, industry analysts predicted that Women.com Networks, Inc. would be a key Internet player for years to come.

Barely more than a year later, everything fell apart.

Marleen audibly exhales when I ask about women.com's fate. "Not my favorite topic," she concedes. "I'd never in my life been in a company where revenue actually declined in a quarter," she says. "I wasn't alone, but I had to report that as a public company. Our stock started sliding, and I could see the handwriting on the wall." She initiated an idea to her board: in order to become un-fucked, women.com needed a merger. "Once you're hurt by the markets, you become one of the living dead," she says. "I moved quickly."

On February 5, 2001, women.com made an unexpected announcement: the network's biggest competitor, iVillage, would acquire women .com in a complicated stock deal valued at forty-seven million dollars. While the merger was still pending, iVillage received a notice from NASDAQ saying that their company stock was teeteringly low, on the brink of being delisted, and almost immediately after the merger went through, the newly doubled company fired half its staff. The *Wall Street Journal* drew an unflinching conclusion: "The Internet cannot sustain several major ad-dependent sites catering to the niche market of women's interest."

Things may not have been so dire. After all, plenty of ad-supported women's sites and apps thrive online today, serving puff content, incisive feminist commentary, and everything in between. The difference between these and the women's sites of the early Web is that the latter imagined themselves as portals: one-stop shops where women could get all the Internet they needed. This might have worked in the very beginning, when women were still in the minority online, but as the Web grew up, so did its users. After a while, a woman's portal began to seem like a misuse of the medium. Why would a woman interested in finance need to surf women.com's money pages when she could just go directly to Bloomberg? Who needed fashion tips from iVillage, for that matter, if *Vogue* had a Web site? It's not that the market couldn't support more than one Web site for women. It's that women, once they got online, could take it from there.

Chapter Thirteen
THE GIRL GAMERS

Computers have no gender, naturally. But the collective social understanding of who computers are for—who uses them, builds them, or buys them—shifts from generation to generation. In the nineteenth century, computers were *actually* women, and in the 1950s, they were a woman's game, until programming was professionalized and masculinized. The ARPANET, built on a technical backbone radiating from military and academic centers, was male dominant because the people designing that infrastructure came from environments that favored men. Early hypertext was designed by women on the peripheries of computer science, but it took a man to popularize their ideas—until finally, in the early years of the Web, access to personal computers let women back in, and a generation of female culture workers and entrepreneurs made their impact yet again. In all that time, the technology has always reflected what was put into it, and by whom.

As the researcher Jane Margolis notes, "Very early in life, computing is claimed as male territory. This claiming is largely the work of a culture and society that links interest and success with computers to boys and men," leading to the cumulative discouragement of women, what Margolis evocatively calls the "bitter fruit of many external influences." This discouragement permeates technical culture at every level:

in school computer labs, in the way a home computer might be placed in a young boy's room rather than in his sister's, in a mass-market media that lionizes male geeks, and even in computer operating systems themselves, which the sociologist Sherry Turkle points out are so brusque they speak of *crashes* and *executions,* and go so far as to ask if we would like to *abort* ailing programs. It also starts young, with the earliest stirrings of a child's gender socialization.

In the mid to late 1990s, while the women of the World Wide Web were fiercely staking their place in a rapidly evolving information ecosystem, another battleground for the hearts and minds of female computer users began to emerge. It wasn't online, and it wasn't in academia, either, although it would have an indelible effect on both. Rather, it was a fight to reshape the very first impressions that computers make on girls, and to redirect their passions toward the screen with an activity that justifies access and representation alike: playing computer games.

COMBAT EPISTEMOLOGIST

Brenda Laurel tells two different stories about the first time she saw a computer.

In the first one, she's twelve. It's Halloween in Ohio, 1962. Describing her costume—an ear of corn—she says, drily, "It was the Andy Warhol period." Her mom built it from a chicken-wire frame stuffed with kernels of yellow cotton. Brenda could hardly see out the eyeholes, but she was convinced she'd earn first prize in a costume contest held at the local supermarket. By the time she'd waddled over, however, another kid had won. Brenda's mother, a "little-bitty feisty woman who thought the world was pointed at her like a gun," complained to the man in charge, the manager of a hardware store. As a consolation, he handed Brenda a prize from off the shelf: a little plastic box with the word "ENIAC" stamped on it. "It's a computer," he said.

The toy ENIAC came with a stack of cards, each printed with a question. He took one from the top of the stack—What is the distance between the Earth and the moon?—placed it in the machine, and turned a crank on the side. The answer spat out of the other end: it was printed on the reverse. Brenda didn't catch the trick, and she had no way of knowing that the real ENIAC would have come up with its answer with the help of a group of women programmers. But she marveled at the idea of a machine in relation to the cosmos. "I had an epiphany," she remembered. "For a moment I was transported out of my chicken-wire cage, out of the age of schoolbooks and typewriters, and into a glorious time when computers would answer the really hard questions for us."

In the second story, she's older. It's the mid-1970s and she's getting her PhD in theater at Ohio State, a fitting pursuit for the would-be winner of a costume contest. One of her closest friends, Joe Miller, has a job at a nearby research lab, and one night he brings her in, after hours, to show her the second computer that she ever saw. It was painting, as she describes it, "pixels from Mars." Just as she'd imagined about the toy ENIAC of her childhood, this was a machine that spoke to the stars. She fell to her knees. "Whatever this is," she said, "I want a piece of it."

Fortunately for Brenda, Joe started a small software company, the kind that existed before the monopolies of Apple and Microsoft. His shop wrote programs for the CyberVision, a primitive home computer system sold exclusively at Montgomery Ward that has been all but forgotten. It plugged into a normal television set—the existing television remote control played double duty as mouse—and its programs came on standard stereo cassettes, or "Cybersettes," running audio and data on separate channels. As a personal computer, the CyberVision was wildly ahead of its time. It offered a whole suite of programs: home finance software, color games, educational tools, and animated fairy tales for children, all rendered in blocky pixels on 2K of RAM.

This last piece was Brenda's first foray into the art of software. She'd done fairy tales in theater, and her wandering production of Robin Hood's tales, staged at intervals among the oak trees of the Ohio State campus, was a hit with children and adults alike. Despite Brenda's lack of programming knowledge, Joe asked her to come work for him, to design CyberVision's pixel fables. It became her education. "Without knowing it was hard," she immersed herself in the world of computers, doing "everything from graphic design to programming to making coffee." In those days, code was hand diagrammed with pencil and paper before being hand converted into the assembly language understood by the CyberVision's CDP1802 microprocessor, an integrated chip used in a number of hobbyist and consumer-level microcomputers in the 1970s. As she squeezed Goldilocks and the three bears into four-color pixel animations, Brenda showed herself the ropes.

Placed prominently in the 1978 Montgomery Ward spring catalog, the original CyberVision sold ten thousand units in its first year, not bad for a computer company from Columbus, Ohio. But the market for personal computers was small, and the competition steep: Sears had the Atari systems, Radio Shack was pushing its own Tandy computer, and the golden age of arcade games was well underway. When CyberVision folded in 1979, Brenda still hadn't finished her dissertation. But no matter—she was a game designer now. When she moved out to California to work for Atari, she saw the ocean for the first time.

We sit in her garden and talk over iced tea; Tejava and pomegranate juice, the house drink. She sucks on an American Spirit, sitting unnaturally straight. She's recovering from back surgery, her second, and her shock of curly silver hair is accented with pops of magenta and aquamarine. As we talk, two muscular cats with Japanese names clamber around the garden, which is really a forest: five acres of pristine land near the Portola Redwoods, a senselessly beautiful two hours' drive from San Francisco. As the sun moves across the picnic bench, it strikes a pile of abalone shells tossed in the sawdust behind her. Brenda and her husband are abalone people; they free dive off the Northern California coast to wrench the opalescent shells from rock faces with a pry bar. The trick, she tells me, is to catch them unaware. Even bivalves lock down when they sense the current of change. It took Brenda four years to pull her first abalone. One day, mid-dive, she remembered that she'd been born left-handed. She switched the pry bar from right to left and pried that sucker in one brute pull, *ka-chunk*. None of these things are metaphors for how Brenda works, but they do add up to something: a woman with stealth and force when it's needed, with a hidden prying arm and a garden full of rainbows, each of them a death.

In California, Brenda strategized software for the Atari 400 and 800 computers. Since Atari had originally made its name in the arcade, the company wanted its new personal computers to run its most iconic games, and Brenda spent a lot of her time and energy porting Atari titles over to the computer. It drove her crazy: the way she saw it, the games already ran better on a console that cost a tenth as much. By the time her team made it to *Ms. Pac-Man,* she went to the president of the computer division and said, "You know what? I can't stand it anymore. Let me show you." She pulled out the whiteboard and listed all the things she thought Atari should be doing in home computers: personal finance, education, word processors. "And the guy said, 'Your salary's doubled and you're reporting to me.'"

These are the kinds of leaps Brenda makes: with both feet. When she decided the department's corporate overseers were gunning for her job, she took herself across the street to Atari's research lab. It was headed by Alan Kay, a computer scientist known for his pioneering work on object-oriented programming and for designing the overlapping windows of a computer desktop. Kay took Brenda under his wing, buying her several more years at Atari. In Kay's lab, she designed an artificial intelligence system, based on Aristotle's *Poetics,* to generate compelling new scenarios for computer games. From there, she jumped to Activision, where she produced games like *Maniac Mansion,* and then Apple, where she brought some of her more radical friends—like LSD pioneer Timothy Leary—into dialogue with the human-computer interface group. Somewhere along the line, she finally finished her dissertation, which argued that computer programs are like theater: they both have scripts, and neither perform, or are performed, the same way twice.

Brenda's house is several miles down a one-lane road through a madrone wood. There's a labyrinth chalked onto the driveway and a shelf of classic *Star Trek* memorabilia in the living room. On her office door, a plaque reads BRENDA LAUREL, PH.D., COMBAT EPISTEMOLOGIST, and in a low closet underneath the stairs, she keeps baskets of fabric, long spools of maypole ribbon, and plastic flowers, their green wire stems akimbo. We have to pull it all aside to get to the plastic tub in the back, stuffed with memorabilia of a different kind. She's promised me we can go through the archives.

Everything in the box is purple. There are toy figurines, vacuum-packed in clamshell plastic. There are sets of trading cards, collections of opalescent beads in purple velvet baggies, and CD-ROMs in purple boxes with names like *Rockett's Tricky Decision* and *Rockett's Adventure Maker,* a couple of the titles she produced when she was the head of her own computer game company, Purple Moon. As she's jamming loose ribbon and tulle back into the cupboard, I read the company's mission statement, printed on the back of a disk: *Deep friendships. Love of nature. The confidence to be cool. The courage to dream. It's what girls are all about. And it's what girls share when they discover Purple Moon adventures. Which is why Purple Moon is just for girls.*

PURPLE MOON

In 1992, Brenda got a job at Interval Research, a Palo Alto think tank funded by Microsoft cofounder Paul Allen. Interval was all *R* and very little *D:* researchers wandered into the weeds studying technologies that were still far from being commonplace, like telepresence and interactive video. Brenda had just come in from the weeds herself. She'd founded a virtual reality company, Telepresence Research, that had folded within a year. She often calls herself a member of the "Crash Test Dummy Club," those breakneck dreamers who try to make things before they're economically feasible. It's an "uncomfortable but fine, wild ride," she says. Painful as the consequences may be for them, crash test dummies always see what's coming up ahead.

Brenda noticed something while trying to develop virtual reality: men and women seemed to experience it differently. "When I interviewed men about VR," she tells me, "they would typically say it was an out-of-body experience." But when she talked to women, "they would typically say they were taking their sensorium into a different environment." It was a nuance, but it was enough to get Brenda thinking about gender and technology and how small imbalances in how technologies are designed have a larger net effect on who uses them, who benefits from them, and who profits. At Interval, she narrowed her focus further, opting to study the generation of children just then coming of age in a personal-computer-driven world—children like her own daughters. During her first four years at Interval, Brenda asked a small question with big implications: Why didn't little girls play computer games?

Games provide many kids with their first hands-on exposure to computers. But just as Brenda was starting at Interval Research, among fourth- through sixth-grade students, heavy users of computers were overwhelmingly male. Researchers at that time found that while girls tended to see the computer as a means to accomplishing a task, like word processing, boys were more likely to "play games, to program, and to see the computer as a playful recreational toy," behavior that breeds familiarity, then mastery. This trend toward socializing computers as the pursuit of male nerds emerged from the long, slow masculinization of software engineering, and it continued to be perpetuated in popular culture, in films like *War Games, Revenge of the Nerds,* and *Weird Science,* in which nebbish boys "program" their dream woman, and from the marketing of computers and computer games throughout the 1980s and 1990s.

After interviewing nearly one thousand children and five hundred adults all over the United States over the course of four years, Brenda came to believe that the problem wasn't about access, or even, really, the representation of computers in media. It didn't add up: there were plenty of girls with computers at school or at home who still weren't using them, and there was no research to substantiate a claim that girls were somehow *inherently* less skilled or interested in computer play. For Brenda, it came down to a software issue. Little girls didn't play computer games because the computer games *were all for boys.*

The girls Brenda and her team interviewed from 1992 to 1996 didn't mince their feelings about the games they had played: they hated dying over and over. Violence stressed them out. And they weren't fans of the way games emphasized mastery, like defeating a difficult final boss or navigating fast-moving terrain without getting killed. "Mastery for its own sake is not very good social currency for a girl," Brenda explains. "They demand an experiential path." Instead of trying again and again to beat a level, a baddie, or the clock, the girls Brenda interviewed preferred to wander around, exploring a virtual world and learning the relationships between its characters and places. In *Life on the Screen,* published only a few years after Brenda began her research, the sociologist Sherry Turkle argued that while men generally see computers as a challenge—something to master and dominate—women see computers as tools, objects to be collaborated with. This "soft mastery," she explained, requires a closeness, a connection to computers that's more like the relationship a musician has with her instrument: intimate, dialogue-driven, relational.

Just as musicians harmonize, Brenda found that her subjects played together regardless of whether the games they played were designed for multiple players. She concluded that girls are naturally collaborative, and that their social experience of play is often as important as the objective of the game itself. Her subjects liked puzzles, discovery, and the immersive conversation with the machine that happened when story lines were strong enough to ensnare them, and they preferred to share their experiences with one another. At that age, I certainly did: after beating the CD-ROM game *Myst* in 1993, I remember jumping up and down on the rug with my best friends.

Now this was *R* that could lead to some *D*. In 1996, Interval spun Brenda's research team into its own company, Purple Moon, which would produce games exclusively for young girls. It stood to reason that if boys were hogging the machines at the school computer lab to play games that girls didn't like, girls would later be disadvantaged in a workplace, and a world, where computer literacy is not only beneficial but necessary. Making games that girls *did* like seemed like the obvious solution. As one female game designer put it, "We cannot expect women to excel in technology tomorrow if we don't encourage girls to have fun with technology today." It was a smart business move, too: girls represented a huge untapped market, and the prevailing wisdom was that anyone who made a computer game that really appealed to them could conceivably double the games industry.

This thinking was likely influenced by the unexpected success that year of *Barbie Fashion Designer,* a Mattel CD-ROM game that allowed its players to dress a virtual Barbie. The first best-selling "girl game," it sold six hundred thousand units in 1996, far outselling first-person shooters like *Doom* and *Quake* and surprising just about everyone in the games industry.

Barbie Fashion Designer illuminated a market for girl games, and it bridged a gap between girls' real-world and computer play; Barbie's digital ensembles could be printed, cut, and glued to fit a real Barbie. In terms of game mechanics, however, it hardly pushed the envelope. The idea of dressing a character wasn't even new: "boy" games offered dress up, too, in the form of customizing an avatar, and that's usually before the actual adventure begins. This is emblematic of how the industry approached girl games before Purple Moon. Designers would often package re-skinned, easier versions of established game mechanics. *Barbie Fashion Designer*'s predecessor, a 1991 *Barbie* game for Super Nintendo, had done much the same thing, replacing coins with perfume bottles and monsters with beach balls. It was a boy game in a dress, what Brenda disdainfully calls "the computer game equivalent of pink Legos."

Purple Moon took a different approach. Rather than softening the edges around "boy" games, Brenda's company doubled down on complex characters and story building. "I mean, it's not only that the characters are lame in most boys' games," she explained. "It's that they're so lame you can't even make up an interesting story about them." Purple Moon produced two series of story games about an eighth-grade girl, Rockett Movado, and her circle of friends at Whispering Pines Junior High. The Rockett games have no levels or repetitive trials, no clocks or leaderboards. They can't even be won. Brenda compares them to an emotional rehearsal space. In *Rockett's New School,* Rockett faces situations familiar to any teen or preteen girl: she has to make friends, navigate tricky social situations, and decide what kind of person she is. Will she invite nerdy Mavis to her party? Will she read another girl's private diary? Will she try to fit in with the popular kids—a clique called "The Ones"—or will she speak her mind when she sees somebody being bullied?

These questions are answered at branching points. Walking into homeroom on the first day of school, Rockett must decide who she will be; the choice is made inside her head, where three different versions of Rockett share a snippet of their inner dialogue. Each represents a path in the story. The terrified Rockett branches the game into a different story line from the can-do Rockett, who ends up bickering with an alpha girl over a coveted back row seat. It's a hypertext choose-your-own-adventure for social development, a first day of school you can do over and over again. One avid Purple Moon gamer, who played Rockett during her formative years, remembers how the games influenced her own development. "I can definitely remember times when, instead of being quick to respond to someone, I thought about my words and what possible repercussions would be for each option I had," she explains. "They truly helped teach me how to socialize and how to get along with people."

These social lessons weren't only learned in Whispering Pines. Purple Moon's Web site, an early social network, extended the world of the Rockett games online, allowing girl gamers to learn more about their favorite characters and meet their fellow real-life players. Although Interval's founder, when first shown a prototype, asked only "Can you do this for boys?," the Purple Moon Web site became a girl universe, what we'd now call a *fandom*. Girls submitted articles to the Whispering Pines school newspaper, inventing their own stories surrounding the plotlines they knew well, hung out in the Rockett forums, and exchanged virtual treasures relating to the game world. And unlike the greater Web, the Purple Moon site was safe, because girls registered their accounts through their parents, and the site had a built-in panic button. "Simple," Brenda explains, with characteristic brio. "Some shit's going on, you push the panic button, you get a screen capture, everything goes to us, we see who was an asshole, we call their parents, we give them a warning." The Purple Moon Web site was arguably as influential as its games: in 1997, it was outdoing Disney's Web site in both hits and dwell time, and it had formed the cornerstone of many young girls' online world.

Purple Moon published a companion series to Rockett's junior high adventures, *Secret Paths*. Where the Rockett games emphasized social skills, the *Secret Paths* games were "Friendship Adventures," and unlike the high-stakes environment of middle school, were soothing and private. *Secret Paths in the Woods,* for example, begins in a tree house. Depending on user preference, outside the window is an impressionistic mountain, a babbling stream ringed with flowers, or an ocher vista, dotted with wild horses. In the room is a "Friendship Box" full of girls who need help. If you press your heart to one, she will reveal her secrets. Miko fears that other girls won't like her because she's too smart; Dana is "really bummed" that her team lost the state soccer championship; Whitney has problems with her stepmother. The objective of *Secret Paths in the Forest* is to help each girl by traveling down her "path," weaving through whispering branches and glowing sunsets. Along the way, puzzles reward players with gemstones representing different strengths the girl might need to overcome her problems. They become a gift—a necklace of amulets, shimmering with the promise of self-reliance.

Secret Paths in the Forest, and later *Secret Paths to the Sea,* satisfied different emotional needs than *Rockett's New School.* They were quiet, slow, meditative games; they taught girls how to "find emotional resources within themselves and how to observe and respond to others' unarticulated needs," in the words of one critic. They were meant to be played alone. And while the Rockett games' emotional navigation gave girls a chance to try out social behaviors without embarrassing themselves, the accompanying series helped them to understand core values of empathy, kindness, and solidarity among girls to anchor them through the changes ahead. The shy young kids of *Secret Paths* become the self-assured preteens of *Rockett,* and there's no right way for them to do it.

There are more generous paths, but Purple Moon's games reflected real life. Kids could play Rockett as sugar-sweet, but sometimes a bad decision leads Rockett to a good place—like hiding out in the music room, where the cutest boy in school, Ruben, serenades her on the piano. This is what got Brenda in trouble.

Purple Moon released six Rockett games. The first two broke the top 100 best-selling games the year they were released, beating industry titans like *John Madden Football.* Like the women's Web, they represented a sea change in the gender dynamics of computer use: by the time Purple Moon established itself, so-called "girl games" were everywhere. Mattel, riding the success of *Barbie Fashion Designer,* released several more Barbie Interactive titles—an ocean adventure, a game for designing and printing party invitations, and *Barbie Magic Hair Styler,* which requires little explanation—while Sega competed with *Cosmo Virtual Makeover* and The Learning Company produced an *American Girl* series. The game designer Theresa Duncan found success with the dreamy, alt-rock story games *Chop Suey, Zero Zero,* and *Smarty.* Her Interactive, an offshoot of American Laser Games, produced another high school adventure, *McKenzie & Co.,* which was like a Rockett game with live-action cut scenes, and a popular series of Nancy Drew games. This critical mass had an amplifying effect. The more studios jumped into girl games, the larger the shelf grew in the store. To kids, that real estate is a whole world; the Purple Moon gamer I spoke with, who was in elementary school during the upwelling, told me she just assumed, as a child, that the computer aisles had *always* been full of games for girls.

The girl games movement coincided with an uptick in female-fronted businesses throughout the 1990s. Like Purple Moon, several of the major girl games studios were woman owned and largely woman staffed, paralleling a larger national trend for entrepreneurial feminism. This made it a hot topic in its day. Popular media found it fascinating: *Nightline* ran a special on girl gaming, and *Time* magazine, in a profile of Purple Moon, joked about girls needing a "rom of their own." It provoked even more chatter in game industry and academic circles, and several scholarly anthologies have been published on the subject. The excitement was largely based on the assumption that girl games would create a "'virtuous cycle' where girls playing computer games would lead to women writing game software, and hence more girl-friendly game experiences, and even more girl gamers." But by 2000, the girl game movement was effectively over, with its most recognizable studios—Purple Moon among them—dead and buried.

After *Barbie Fashion Designer,* no girl game studio managed to produce a blockbuster title. Industry observers eventually wrote the game off as a fluke, crediting its sales to Mattel's brand recognition: the typical American girl at the time owned, on average, nine Barbies. But many critics argued that the fundamental thinking behind girl games was flawed, and that designing games explicitly for girls actually undermined them by forcing them into pink-and-purple fishbowls. Instead of separating kids by gender, critics argued, why not make games that all kids could enjoy? The first really successful computer games, like *Pong* and *Tetris,* had no explicit gender, and it wasn't until Atari put a pink bow on Pac-Man to create *Ms. Pac-Man* in 1980 that the issue was even raised.

Brenda's research led her to believe that only boys enjoyed shooters and adventure games, but many girls, then and now, love blasting aliens and annihilating bosses with sprays of machine-gun fire. As one self-professed "Game Grrl" wrote in the late 1990s, "What Purple Moon and other 'girlie games' companies have to understand is that although there is a market for games like *Barbie Fashion Designer,* there is just as big a market for girls who like to do the same things boys do."

Purple Moon found itself embroiled in a much larger cultural conversation about the influence of games on kids' lives, and its girl-forward approach was read as essentialism. In an influential editorial, the feminist critic Rebecca Eisenberg questioned the research that spawned Purple Moon, arguing that the Rockett games, with their focus on "popularity and fashion," only "reinforce the very same stereotypes they purport to combat." While Brenda Laurel designed Whispering Pines Junior High to mirror the social realities of the girls she'd interviewed, Purple Moon's critics lambasted the company for presenting too narrow a view of girlhood and for emphasizing the same patterns of exclusion and affiliation feminists work so hard to deconstruct. Didn't girls deserve better than a cartoon version of a world they already knew? When I ask Brenda about this, she's more pragmatic. She was making an intervention at the level of popular culture, and that required a certain broad appeal. "You can't get buy-in from somebody when they don't recognize the character they're playing," she says.

In 1999, Interval pulled the plug on Purple Moon. The company did not have a graceful end. "Paul Allen took us into Chapter 7," Brenda says, "and we had to talk him out of that and get back into Chapter 11 so we could sell the damn thing to Mattel at a bargain-basement price and make everybody's paycheck. It was just hideous, how it came out." Brenda was forced to shut down the Purple Moon Web site, where girls from all over the world had formed intense friendships, and to fire most of her staff—80 percent of whom were women, the inverse of most Silicon Valley workplaces—in a day. At a game developer's conference in San Jose that year, she blamed Purple Moon's failure on a tech marketplace that valued short-term gains over long-term progress and which fetishized incorporeal Web businesses over those actually selling physical, warehoused goods. "I promised to talk to you about some approaches to research and design that will help us grow both our audiences and our ideas," she told the crowd, according to the *New York Times*. "But since Purple Moon did not make it to the big I.P.O. or a lavish acquisition that made everybody rich, it would be understandable for you to conclude the methods I'm advocating for don't necessarily work." She later published a book titled *Utopian Entrepreneur*.

Barbie bought Rockett on a shopping spree: in 1999, when Mattel acquired Purple Moon, it also picked up The Learning Company, which produced the *Carmen San Diego* games, and Pleasant Company, known for its American Girl dolls and games, before shutting down their games group entirely, effectively kneecapping the girl games movement. "Mattel was trying to protect their Barbie franchise," Brenda says. "They bought everybody and drove a stake through their hearts." But by the late 1990s, the most recognizable female figure in video games wasn't

Rockett or even Barbie: it was the buxom *Tomb Raider,* Lara Croft, with a hypersexualized body and skin-tight ensemble that earned her legions of male fans.

After Purple Moon folded, Brenda held a wake at her house. Former employees and friends of the company gathered around a toy Rockett figurine, laid to rest in a miniature plastic locker on Brenda's antique dining room table, flanked by "black candles, bouquets of purple irises, and a seriously depleted bottle of Irish whiskey." Brenda poured herself a shot and gave a eulogy to the little red-haired girl on which she'd staked the future of games. "We're always trying to heal something," she said. "Lousy childhoods, raw deals, crappy self-esteem. We were trying to heal something when we made her." Brenda has three daughters, and it's clear her mission was personal. They were preteens when Purple Moon shipped its first game, and barely a few years older when the company was shut down. Like any passage through the final stage of childhood, it wasn't long enough. To console their mother, the girls strung the Rockett figurine on a length of pink yarn and dangled her from the second-floor landing like an angel.

Before I leave her house, Brenda runs into her office and plucks something from an altar in the corner: a milky blue bead, shaped like a moon. It's one of the original *Secret Paths* gems, sold by Purple Moon in packs of six, in little purple velvet baggies, for girls to collect and trade. She hands it to me triumphantly. On the back is inscribed a single word: maturity. It's presumably the trait I need to complete the necklace of strength that will help me through the next phase of my life. It seems incongruous for a plastic toy to tell anyone to grow up. But I look at Brenda, at the altar of sacred objects at her side, and I realize: it's not *telling* me to grow up. It's trying to teach me how.

The gem comes home with me, down the long path through the forest.

Epilogue

THE CYBERFEMINISTS

In the summer of 1992, a billboard popped up overnight on the side of an art gallery in Sydney, Australia. Eighteen feet long, it featured a strange assortment of images: spherical fragments of DNA, vaginal wedges of color, and a pair of mirrored women with unicorn horns, flexing their muscles and emerging from seashells. In the middle was a blob of text, rendered convex as a bead of water. WE ARE THE VIRUS OF A NEW WORLD DISORDER, it read, DISRUPTING THE SYMBOLIC FROM WITHIN. SABOTEURS OF BIG DADDY MAINFRAME. THE CLITORIS IS A DIRECT LINE TO THE MATRIX.

Not a moment too soon, the Cyberfeminists had arrived.

The billboard, an artwork called *A Cyberfeminist Manifesto for the 21st Century,* was produced by a four-woman art collective from Adelaide: Josephine Starrs, Julianne Pierce, Francesca da Rimini, and Virginia Barratt, known together as VNS Matrix. Their "blasphemous text" was written in one night a few years previous, as a free-association about "new representations of women, gender, and sexuality in technospace, both primordial, ancient and futuristic, fantastic and active," as Barratt explained to me several years ago. In 1991, VNS Matrix wheatpasted their manifesto onto city walls and faxed it to tech magazines and feminist artists around the world, proclaiming the dawn of a new age: a century and a half after Ada Lovelace first scratched a computer program onto paper, it was time for women to become the virus, the signal, and the pulse of the network.

When the manifesto took billboard form, a student from Britain photographed it and brought the photo home to her professor, the cultural theorist Sadie Plant, who was designing a curriculum in a similar spirit. In her 1997 book, *Zeroes + Ones,* Plant explains that when VNS Matrix wrote "the clitoris is a direct line to the matrix," they meant both the womb (*matrix* is Latin) and "the abstract networks of communication . . . increasingly assembling themselves" in the world around them. It was an evocative vision of women's bodily connection to networked computing, a connection that emerged before the technology itself, beginning with Ada Lovelace and the countless uncounted female computers—a lineage Plant traces in her book, much as I have in mine.

Sadie Plant and VNS Matrix are considered the matriarchs of cyberfeminism, a wild, breathlessly utopian, very brief art movement that flourished in the mid-1990s, as the Web began to reshape the world. Cyberfeminism conjures, in many ways, the countercultural, techno-utopian feeling of early Internet culture, and inherits the spirit of those West Coast cyberhippies who believed that computer-mediated communication would create a free civilization of the mind. The motley crew of artists, coders, game designers, and writers who pronounced themselves cyberfeminists joyfully subverted what VNS Matrix called "big daddy mainframe": the patriarchy hard coded to the technological underpinnings of the world, a backbone built by men. "The technological landscape was very dry, Cartesian, reverent," Barratt says. "It was uncritical and overwhelmingly male-dominated. It was a masculinist space, coded as such, and the gatekeepers of the code maintained control of the productions of technology."

After so many generations of women's technological accomplishments being buried by time, indifference, and the shifting protocols of the network itself, the cyberfeminists were hungry to claim their place in the technological now, and loudly. Cyberfeminist thinkers and artists understood the Internet as an unprecedented platform for free thought and expression, like a dormant virus in the mainframe. The prefix "cyber" summoned it. Ubiquitous at the time—with cyberculture, cyberdelics, cybersex, cyberpunk, and CyberSlacker, too, of course—"cyber" evoked the collective hallucination of digital space and the placeless, incorporeal world of electronic networks. The cyberfeminists were fascinated by the idea of online space without geography, without predefined conventions, and believed a new kind of feminism might set sail there, afloat and untethered on an ocean of fiber and bits. "The Internet was far less regulated, far less commodified," VNS Matrix's Francesca da Rimini says. "More of a maul and a maw than a mall. There seemed to be endless possibilities."

To the many women coming online in the early 1990s, cyberfeminism looked for all the world like the next big wave of feminism: if the previous generation had thought globally but acted locally, holding their consciousness-raising sessions in each other's living rooms, then the Internet could collapse the difference, creating a global living room where pixels and code did the work of pickets and fists.

Indeed, the first generation of feminists to the Web understood that access was an equality issue, and they learned to translate the organizing and campaigning experience of second-wave feminism to the new medium. Some of their earliest efforts on the Web include informational Web sites for victims of abuse, feminist forums, and an animated GIF candlelight vigil for domestic violence. "As the population becomes widely familiar with the new communications technologies," wrote Scarlet Pollock and Jo Sutton, editors of the Canadian feminist magazine *Women'space,* "the challenge to feminists is whether we'll be online and ready to greet them."

Cyberfeminist artists made revolutionary CD-ROMs, created Web-based multimedia artworks, and built virtual worlds, taking many forms as they swam through the network seeking pleasure and knowledge. They wrote howling agitprop like the *Cyberfeminist Manifesto for the 21st Century.* They formed coalitions, mailing lists, and discussion groups, like the Old Boys' Network, a group that proclaimed cyberfeminism to be, above all, "a question of survival and power and fun." VNS

Matrix even made a video game, *All New Gen,* in which the player must hack into the databanks of Big Daddy Mainframe, the Oedipal embodiment of the techno-industrial complex, and slime him and his cohort ("Circuit Boy, Streetfighter, and other total dicks") into oblivion, sowing the seeds of the New World Disorder and ending the rule of phallic power on Earth.

Like the network itself, the movement was expansive and multifarious. "Cyberfeminism only exists in the plural," pronounced the Swiss art critic Yvonne Volkart in 1999. Even at the height of the word's usage, it could never be trusted to mean any single, specific approach to feminism at the dawn of the Web revolution. Instead, the word "cyberfeminism" granted currency to an array of positions, some of which were mutually exclusive. At the First Cyberfeminist International, a 1997 gathering in Kassel, Germany, attendees settled against defining the term, instead collectively authoring one hundred "Anti-Theses," a laundry list of things cyberfeminism was *not*. That list includes the following: not for sale, not postmodern, not a fashion statement, not a picnic, not a media hoax, not nice, not Lacanian, not science fiction, and—my personal favorite—"not about boring toys for boring boys."

Privately, however, the cyberfeminists worried that what was being hailed as "the virtual techno-paradise of the new millennium," as the social scientist Renate Klein wrote in 1999, might eventually become as "woman-hating as . . . much of real life at the end of the twentieth century." To cut off that eventuality, they hoped to move quickly, establishing an online presence colorful, confident, and vivid enough to create a permanent association between women and technological culture, which is their birthright.

But being a woman online today comes with the same anxieties that have always followed women and other marginalized groups, and fears of being silenced, excluded, and bullied remain as palpably real in the digital realm as they are IRL. Our dense net of connective technologies, and the increasing facility by which we are surveilled within them, has led to new forms of violence: doxxing, cyberstalking, trolling, revenge porn. And anonymity, which the cyberfeminists, along with many early cyberculture thinkers, championed as a method for transcending gender and difference, enables violently misogynistic language all over the Web: in YouTube comments, on forums, on Reddit and 4chan, and in the in-boxes and @replies of women with public opinions. The incorporeal newness that so intoxicated the earliest women online has morphed; it has become what the games critic Katherine Cross aptly calls a "Möbius strip of reality and unreality," in which Internet culture "becomes real when it is convenient and unreal when it is not; real enough to hurt people in, unreal enough to justify doing so."

As a movement, cyberfeminism disappeared with the popping of the dot-com bubble. "We did what we had to do at the time," Barratt explains. "Our job as female-identified people, and as feminists, was to overthrow the gatekeepers in order to access a powerful new technology which had huge implications for domination and control by the patriarchy and by capitalist systems." As the Web commercialized, it became clear that the Internet was *not* going to liberate anyone from sexism, or for that matter from divisions of class, race, ability, and age. Instead, it often perpetuated the same patterns and dynamics of the meatspace world. Capitalist systems have won, the personal brand is king, and, as ongoing battles for net neutrality have revealed, the gatekeepers still hold tightly to the keys.

It's not that the cyberfeminists, or any of their predecessors, have failed. It's that as digital and real life edge into near-complete overlap, the digital world inherits the problems of the real. Trace a pen across the Möbius strip and it leads you right back to where you started. In this continuous surface, it's harder and harder to draw distinctions. The computers are smaller now, and they come with us to bed; they measure our breathing as we sleep; they listen and track us as we navigate the world. Social networks have built empires by selling us what we already want, and our opinions are formed in bubbles, in a continuous loop of algorithmic feedback. For better or worse, we have become the network, bodies and all.

But that can be a good thing, too. Because as we map our society ever more closely to the screen, we create an increasingly powerful tool for changing it. Lies online can become truth if they are propagated widely enough, and social media has remade the way we travel, eat, and start revolutions: every decision made in the design of our most intimate technologies affects our lives, our cities, our social structures, and our collective experience of what is right, real, and true. When we create technologies, we don't just mirror the world. We actually make it. And we can remake it, so long as we understand the awesome nature of the responsibility.

The more diversity there is at the table, the more interesting the result onscreen, the more human, as Stacy Horn would say, *bite me,* the better. There's no right kind of engineer, no special plane of thought that must be reached to make a worthwhile contribution. There's no right education, no right career path. Sometimes there isn't even a plan. The Internet is made of people, as it was made for people, and it does what we tell it to do.

We can remake the world.

The first step is to see it clearly, seeing who was really there at the most pivotal points in our technological history, without taking for granted the prevailing myths of garages and riches, of alpha nerds and brogrammers. The second step is to learn all the strategies of triumph and survival we can from our forebears, and I hope this book has unearthed a few: Ada Lovelace's refusal of propriety, Grace Hopper's forward-thinking tenacity, and the support the women of Resource One gave one another. Jake Feinler's clarity of vision in the chaos of a changing network. A draft of Jaime Levy's punk rock spirit for courage, and a healthy helping of VNS Matrix's bodily self-assurance that the Internet is our place, wild and weird and mind-bending, as it has always been.

The final step is the hardest: we get to work.

Acknowledgments

Writing a book is the ultimate life hack; I know of no better way to meet your heroes. I am deeply indebted to the people who opened up their calendars, their hard drives, and their memories for me. Not everyone made it into the final version of this book, but you have all had an immeasurable influence on my thinking: Pat Wilcox, Pamela Hardt-English, Sherry Reson, Joan Lefkowitz, Mya Shone, Chris Macie, Lee Felsenstein, Elizabeth "Jake" Feinler, Mary Stahl, Radia Perlman, Aliza Sherman, Ellen Pack, Nancy Rhine, Naomi Pearce, Stacy Horn, Marisa Bowe, Jaime Levy, Howard Mittelmark, Dame Wendy Hall, Annette Wagner, Cathy Marshall, Judy Malloy, Karen Catlin, Nicole Yankelovich, Gina Garrubbo, Laurie Kretchmar, Marleen McDaniel, Naomi Clark, Brenda Laurel, Adriene Jenik, Amy Bruckman, Antoinette LaFarge, Cynthia DuVal, Helen Varley Jamieson, Judy (yduJ) Anderson, Juli Burk, Lisa Brenneis, Lynn Finch, Pavel Curtis, Jim Bumgardner, and Yib. I've been an awed and grateful witness, but this is only the beginning. There are more fascinating stories about women in technology than I could ever have fit in these pages, and I sincerely hope to read many, many more books on the subject in the years to come.

Thanks are due, also, to those who helped along the way: to Robert Kett and Martina Haidvogl at SFMOMA, who helped me to consult the CD-ROM archive of *Word* magazine in the museum's permanent collection; Wende Cover at the Internet Hall of Fame, who connected me with early networking pioneers; Sydney Gulbronson Olson at the Computer History Museum, who handled my queries about Community Memory; and the saintly people of the Internet Archive, without whose Wayback Machine the dot-com-era chapters would have been impossible to write. Give them all your money. I'm not a trained historian, and I am deeply appreciative of the work done by the scholars of computing history cited throughout this book, particularly in the early chapters. They're doing world-changing work, in many cases righting egregious exclusions. When the document overwhelmed me, I turned to the best writers I know. Brian Merchant shared war stories, software strategies, and essential insights on early chapters and drafts; Corrina Laughlin was always deft with an e-mail full of research leads, PDFs, and feedback. Kathryn Borel Jr. was an invaluable sounding board, and Addie Wagenknecht, the ultimate cyberfeminist, gave the book a thorough technical read.

Without my exceedingly thoughtful and ambitious literary agent, Sarah Levitt, this book would still be a three-page document in a folder labeled "someday." She saw something bigger than I had hoped to imagine, and she brought me there. Thank you to my editor, Stephanie Frerich, for her eagerness to tell women's stories, for helping to pull this book from the weeds, and for backing me up. Thank you to Olivia Peluso and everyone at Portfolio for walking me through this process with so much care; thank you to my copy editor, Juliann Barbato, and to my production editor, Ryan Boyle, for taking time to clarify even the thorniest stuff. Thank you to my parents, Colin and Rosine Evans, for the Dell and for their unending and unquestioning support, and to Jona Bechtolt, who draws me out of myself: thank you for heading fearlessly into the darkest caves with me. You're the light.

Notes

CHAPTER ONE: A COMPUTER WANTED

9 **"A Computer Wanted," it says:** "A Computer Wanted," *New York Times,* May 2, 1892.

10 **original cottage industry:** James Gleick, *The Information: A History, A Theory, A Flood* (New York: Vintage Books, 2012), 84.

10 **offices of his time did "mental labor":** Charles Babbage, *On the Economy of Machinery and Manufactures* (London: Charles Knight, Pall Mall East, 1832), 153.

11 **mathematicians would guesstimate their horsepower:** David Alan Grier, *When Computers Were Human* (Princeton, NJ: Princeton University Press, 2005), 276.

11 **"the visible pattern" of any cloth:** Sadie Plant, *Zeroes + Ones: Digital Women and the New Technoculture* (London: Fourth Estate, 1998), 66.

12 **"half a framebreaker":** George Gordon Byron, *Lord Byron: Selected Letters and Journals,* ed. Leslie A. Marchand (Cambridge, MA: Harvard University Press, 1982), 58.

13 **"It is a known fact," Babbage proclaimed:** Charles Babbage, *Passages from the Life of a Philosopher* (New Brunswick, NJ: Rutgers University Press, 1994), 116–17.

13 **writing of a "store" to hold the numbers:** Ibid., 117.

13 **"very costly toy":** Gleick, *The Information,* 101–5.

14 **"mad, bad, and dangerous to know":** Betty Alexander Toole, *Ada, the Enchantress of Numbers: Prophet of the Computer Age* (Mill Valley, CA: Strawberry Press, 1992), 6.

14 **"Oh, my poor dear child!":** Ibid., 21.

14 **"I do not believe that my father was":** Ibid., 156–57.

16 **"aptitude for grasping the strong points":** B. V. Bowden, "A Brief History of Computation," in *Faster Than Thought: A Symposium on Digital Computing Machines,* ed. B. V. Bowden (London: Pitman and Sons, 1953), 22.

16 **He was an "old monkey":** Toole, *Ada, the Enchantress,* 33.
16 **"While other visitors gazed":** Sophia Elizabeth De Morgan, *Memoir of Augustus De Morgan* (London: Longmans, 1882), 89.
17 **"I hope you are bearing me in mind":** Toole, *Ada, the Enchantress,* 83.
17 **"intuitive perception of hidden things":** Ibid., 101.
20 **"My dear and much admired Interpretress":** Ibid., 172.
20 **"The Analytical Engine *weaves algebraical patterns*":** Ibid.,182.
21 **"That brain of mine":** Ibid., 147.
21 **"He is an uncommonly fine baby":** Ibid., 155.
21 **"Not even countesses":** Plant, *Zeroes + Ones,* 32.
22 **One biographer has suggested:** Benjamin Wooley, *The Bride of Science: Romance, Reason, and Byron's Daughter* (New York: McGraw-Hill, 1999), 340–41.
22 **"I do dread that horrible struggle":** Toole, *Ada, the Enchantress,* 290.
22 **"her ideas are so modern":** B. V. Bowden, preface to *Faster Than Thought,* xi.
23 **the astronomer Edward Charles Pickering:** The going legend here, although there is some evidence to the contrary, is that Pickering hired Fleming after growing frustrated with a group of male assistants hired to inspect photographic plates of stellar spectra. Storming out of his office, he vowed that even his Scottish maid could do a better job. He was more right than he knew.
23 **"The Harvard Computers are mostly women":** Grier, *When Computers Were Human,* 83.
23 **Known to history as "Pickering's Harem":** Gabriele Kass-Simon, *Women of Science: Righting the Record* (Bloomington: Indiana University Press, 1993), 100.
24 **"the talented daughters of loving fathers":** Grier, *When Computers Were Human,* 81.
24 **ballparked a unit of "kilogirl" energy:** Ibid., 276.
24 **kept its own pool of "girls":** Beverly E. Golemba, *Human Computers: The Women in Aeronautical Research* (unpublished manuscript, 1994), 43, https://crgis.ndc.nasa.gov/crgis/images/c/c7/Golemba.pdf.
24 **Of these, the mathematician Katherine Johnson:** Sarah McLennan and Mary Gainer, "When the Computer Wore a Skirt: Langley's Computers, 1935–1970," *NASA News & Notes* 29, no. 1 (2012), https://crgis.ndc.nasa.gov/crgis/images/c/c3/Nltr29-1.pdf.
25 **"when the computer wore a skirt":** Jim Hodges, "She Was a Computer When Computers Wore Skirts," 2008, www.nasa.gov/centers/langley/news/researchernews/rn_kjohnson.html.

CHAPTER TWO: AMAZING GRACE

27 **As Grace began her graduate studies:** From which she was only the eleventh woman to receive a PhD in mathematics.
27 **Her intellectual ambidexterity was legendary:** Kathleen Broome Williams, *Grace Hopper: Admiral of the Cyber Sea* (Annapolis, MD: Naval Institute Press, 2004), 12.
28 **It was a nice break from the breakneck:** Williams, *Grace Hopper,* 16.
28 **"one jump ahead of the students":** Grace Murray Hopper, interview by Uta Merzbach, July 1968, Computer Oral History Collection, Archives Center, Na-

tional Museum of American History, Smithsonian Institution, 16, http://amhistory.si.edu/archives/AC0196.pdf.

28 **it was a "gorgeous year":** Ibid., 28.

29 **"I was beginning to feel pretty isolated":** Ibid., 25.

29 **"We usually ended up going through together":** Ibid.

30 **"I just reveled in it":** Ibid., 26.

31 **"Where have you *been?*":** Ibid., 29.

32 **Everyone at Harvard called it the Mark I computer:** Kurt W. Beyer, *Grace Hopper and the Invention of the Information Age* (Cambridge, MA: MIT Press, 2009), 37.

32 **"That is a computing engine":** Hopper, interview by Merzbach, 1968, 29.

32 **The positions of holes:** Grace became so accustomed to eight-bit coding that she'd sometimes accidentally balance her checkbooks in octal.

33 **"this gray-haired old schoolteacher":** Hopper, interview by Merzbach, 1968, 29.

33 **"And then he gave me a week":** Ibid.

33 **Years later, when Grace was an established figure:** Beyer, *Grace Hopper,* 314.

34 **"It was fascinating," she said, a "hotbed of ideas":** Hopper, interview by Merzbach, 1968, 31.

34 **"bawling out" the perpetrator:** Aiken's reputation was so widely known that even a laudatory article about his retirement in the April 1962 issue of *Communications of the ACM,* a trade journal for the field he helped pioneer, described him as "strong-willed, independent, single-minded and insistent on the highest standards of scholarly integrity, performance, and achievement" and "a ruthless taskmaster."

34 **"He's wired a certain way":** Grace Murray Hopper, interview by Beth Luebbert and Henry Tropp, July 1972, Computer Oral History Collection, Archives Center, National Museum of American History, Smithsonian Institution, 29, http://amhistory.si.edu/archives/AC0196.pdf.

34 **And anyway, as Grace told Howard Aiken:** Ibid., 47.

35 **"pull her mirror out of her pocketbook":** Howard Aiken, interview by Henry Tropp and I. B. Cohen, February 1973, Computer Oral History Collection, Archives Center, National Museum of American History, Smithsonian Institution, 44, http://amhistory.si.edu/archives/AC0196.pdf.

35 **"Grace was a good man":** Ibid.

35 **"Bug" is engineering slang:** Fred R. Shapiro, "Etymology of the Computer Bug: History and Folklore," *American Speech* 62, no. 4 (1987): 376–78.

35 **"gremlin that had a nose":** Hopper, interview by Luebbert and Tropp, 1972, 27.

36 **After the moth incident:** Grace Murray Hopper, interview by Uta Merzbach, January 1969, Computer Oral History Collection, Archives Center, National Museum of American History, Smithsonian Institution, 13, http://amhistory.si.edu/archives/AC0196.pdf.

36 **"ninety-nine percent of the time":** Ibid., 10.

37 **Back before the war:** Williams, *Grace Hopper,* 13.

38 **"The tremendous contrast":** Hopper, interview by Merzbach, 1968, 4.

39 **draftswomen, assemblers, secretaries, and technicians:** Thomas Haigh, Mark Priestly, and Crispin Rope, *ENIAC in Action: Making and Remaking the Modern Computer* (Cambridge, MA: MIT Press, 2016), 298.

39 **"basically *Angry Birds*":** Mark Priestley and Thomas Haigh, *Working on ENIAC: The Lost Labors of the Information Age,* http://opentranscripts.org/transcript/working-on-eniac-lost-labors-information-age.
41 **In their backbreaking calculations:** John Mauchly, interview by Uta Merzbach, June 1973, Computer Oral History Collection, Archives Center, National Museum of American History, Smithsonian Institution, 22, http://amhistory.si.edu/archives/AC0196.pdf.
41 **Dorothy's ability to code:** Jean Jennings Bartik, *Pioneer Programmer: Jean Jennings Bartik and the Computer That Changed the World* (Kirksville, MO: Truman State University Press, 2013), 9.
42 **"more handicraft than science":** Nathan Ensmenger, *The Computer Boys Take Over: Computers, Programmers, and the Politics of Technical Expertise* (Cambridge, MA: MIT Press, 2010), 15.
42 **"We were perplexed and asked":** Bartik, *Pioneer Programmer,* 13.
42 **In its time at the Moore School:** Haigh et al., *ENIAC in Action,* 96–97.
43 **"like an automaton":** John Mauchly, interview by Henry Tropp, January 1973, Computer Oral History Collection, Archives Center, National Museum of American History, Smithsonian Institution, 70, http://amhistory.si.edu/archives/AC0196.pdf.
43 **"It was just a great romance":** Ibid.
43 **"I don't want teaching":** Kay McNulty, quoted in W. Barkley Fritz, "The Women of ENIAC," *IEEE Annals of the History of Computing* 18, no. 3 (Fall 1996): 16.
44 **It had forty panels:** H. H. Goldstine and Adele Goldstine, "The Electronic Numerical Integrator and Computer (ENIAC)," *IEEE Annals of the History of Computing* 18, no. 1 (Spring 1996), 10.
44 **"little IBM maintenance man":** Jean J. Bartik and Frances E. "Betty" Snyder Holberton, interview by Henry Tropp, April 1973, Computer Oral History Collection, Archives Center, National Museum of American History, Smithsonian Institution, 19, http://amhistory.si.edu/archives/AC0196.pdf.
44 **They found a sympathetic man:** Ibid., 21.
46 **"a little cavalier":** Mauchly, interview by Tropp, 1973, 66.
46 **"How do you write down a program?":** Bartik and Holberton, interview by Tropp, 1973, 29.
46 **Occasionally, the six of us programmers:** Fritz, "The Women of ENIAC," 1996, 1096.
47 **"cross between an architect and a construction engineer":** *The Computers: The Remarkable Story of the ENIAC Programmers,* directed by Kathy Kleiman (2016, Vimeo), VOD.
47 **"It was a son of a bitch to program":** Bartik, *Pioneer Programmer,* 84.
47 **"Betty and I had a grand time":** Ibid., 84–85.
48 **The pair worked around the clock:** Ibid., 92.
48 **Betty Jean kept a taste:** Ibid., 95.
49 **Betty could "do more logical reasoning":** Ibid., 85.
49 **faster than a speeding bullet:** Ibid., 25.
49 **The Bettys and Kay McNulty hustled:** Jean Jennings Bartik, "Oral History of Jean Bartik: Interviewed by Gardner Hendrie," July 1, 2008, Computer History Museum, 31, www.computerhistory.org/collections/oralhistories.
50 **"The amount of work that had to be done":** Bartik and Holberton, interview by Tropp, April 1973, 55.

50 **"the ENIAC was... told to solve":** T. R. Kennedy, "Electronic Computer Flashes Answers, May Speed Engineering," *New York Times,* February 15, 1946.

50 **"several weeks' work" would never:** Jennifer S. Light, "When Computers Were Women," *Technology and Culture* 40 (1999): 474.

50 **"The press conference and follow-up":** Ibid.

51 **"It felt like history had been made":** Bartik, *Pioneer Programmer,* 85.

51 **"boldface lie":** Bartik, interview by Gardner Hendrie, July 2008, 31.

51 **"I wasn't photogenic":** Janet Abbate, *Recoding Gender: Women's Changing Participation in Computing* (Cambridge, MA: MIT Press, 2012), 37.

51 **When the army used a War Department:** Light, "When Computers Were Women," 475.

51 **genderless "group of experts":** Ibid., 473.

52 **"If the ENIAC's administrators had known":** Bartik, *Pioneer Programmer,* 21.

52 **"subprofessional, a kind of clerical work":** Jennifer S. Light, "Programming," in *Gender and Technology: A Reader,* ed. Nina Lerman, Ruth Oldenziel, and Arwen P. Mohun (Baltimore, MD: The Johns Hopkins University Press, 2003), 295.

CHAPTER THREE: THE SALAD DAYS

54 **He promptly remarried:** Williams, *Grace Hopper,* 17.

54 **"My time was up":** Hopper, interview by Merzbach, 1969, 15.

55 **If the computer didn't run:** Ibid.

55 **But computers had a huge:** Abbate, *Recoding Gender,* 42.

56 **"I loved it," she wrote:** Bartik, *Pioneer Programmer,* 140.

56 **"The fact is," Betty Snyder said:** Frances E. "Betty" Holberton, interview by James Ross, April 1983, Charles Babbage Institute, Center for Information Processing, University of Minnesota, Minneapolis, 10, www.cbi.umn.edu/oh.

56 **"a very delightful person":** Hopper, interview by Merzbach, 1969, 3.

57 **"We all accepted Pres":** Bartik, *Pioneer Programmer,* 138–40.

57 **A year into her employment:** Ibid., 123.

57 **"That's how so many secretaries":** Captain Grace Hopper, "Oral History of Captain Grace Hopper: Interviewed by Angeline Pantages," December 1980, Computer History Museum, 27, www.computerhistory.org/collections/oralhistories.

58 **But only EMCC had working machines:** Beyer, *Grace Hopper,* 171.

58 **"slipped into UNIVAC like duck soup":** Hopper, interview by Merzbach, 1969, 10.

58 **"if the programmer and the engineer":** Holberton, interview by Ross, 1983, 6–7.

59 **"if anyone could do something":** Bartik, *Pioneer Programmer,* 123.

59 **"Now I had two-dimensional programs":** Hopper, interview by Merzbach, 1969, 3.

60 **"At that time the Establishment":** Grace Murray Hopper, "Keynote Address," in *History of Programming Languages,* ed. Richard L. Wexelblat (New York: Academic Press, 1981), 9.

61 **"When Remington Rand bought UNIVAC":** *UNIVAC Conference, OH 200* (Oral history on May 17–18, 1990, Washington, DC, Charles Babbage Institute, University of Minnesota, Minneapolis, http://purl.umn.edu/104288).

61 **"I mean, it was just as though":** *UNIVAC Conference,* 1990.

61 **"thought these idiots down in Philadelphia":** Ibid.

61 **"That was a disaster":** Holberton, interview by Ross, 1983, 14.

62 **An excerpt from that oral history:** UNIVAC Conference, 1990.
62 **"There was no feeling":** Holberton, interview by Ross, 1983, 14.
63 **"We are at a loss":** Beyer, *Grace Hopper,* 217–18.
63 **"it sounded impressive enough to match":** Hopper, interview by Merzbach, 1968, 8.

CHAPTER FOUR: TOWER OF BABEL

64 **Not coming from any existing art:** Jean Sammet, *Programming Languages: History and Fundamentals* (Englewood Cliffs, NJ: Prentice-Hall, 1969), 44–53.
64 **"guarding skills and mysteries":** John Backus, "Programming in America in the 1950s: Some Personal Impressions," in *A History of Computing in the Twentieth Century,* eds. N. Metropolis, J. Howlett, and Gian-Carlo Rota (New York: Academic Press, 1980), 127.
66 **The most basic programs specify:** Abbate, *Recoding Gender,* 76.
67 **"looking at a DNA molecule":** Douglas Hofstadter, *Gödel, Escher, Bach: An Eternal Golden Braid* (New York: Basic Books, 1979), 290.
67 **Ostensibly, a computer like the UNIVAC:** Like many people in the 1950s, Grace uses "UNIVAC" to mean "computer."
67 **"the novelty of inventing programs":** Grace Hopper, "The Education of a Computer," ACM '52, Proceedings of the 1952 ACM National Meeting, Pittsburgh, 243–49.
67 **"a well-grounded mathematical education":** Ibid.
68 **running print advertisements:** Abbate, *Recoding Gender,* 86.
68 **"It was very stupid":** Hopper, interview with Pantages, 1980, 7.
69 **Grace saw the proliferation:** The biblical metaphor was used by Grace Hopper and stuck. Even the cover of Jean Sammet's canonical *Programming Languages: History and Fundamentals,* the first major analysis of the field, features a winding tower inscribed with the names of a hundred different languages.
69 **"time for a common business language":** R. W. Bemer, "A View of the History of COBOL," *Honeywell Computer Journal* 5, no. 3 (1971): 131.
70 **Every major computer manufacturer:** In 1959, that meant IBM, Honeywell, RCA, General Electric, Burroughs, National Cash Register, Philco, Sylvania, International Computers and Tabulators, and Sperry Rand, the company that resulted from the merger of Remington Rand and Sperry Gyroscope.
70 **The first would examine existing compilers:** Ensmenger, *The Computer Boys,* 94.
70 **"This language was going to be 'it'":** Betty Holberton, "COBOL Session: Transcript of Discussant's Remarks," in *History of Programming Languages,* 262.
70 **beyond Holberton and Mary Hawes:** When Jean Sammet was a young programmer, she worked at the Sperry Gyroscope Company, a defense contractor. In 1955, Sperry bought out Remington Rand, forming Sperry Rand. Sammet would often take the night train to Philadelphia to run programs on UNIVAC computers before they were shipped, serving as a beta-tester in Grace Hopper's programming division. She remained a great admirer of Grace throughout her career. Steve Lohr, *Go To: The Story of the Math Majors, Bridge Players, Engineers, Chess Wizards, Maverick Scientists, and Iconoclasts—the*

Programmers Who Created the Software Revolution (New York: Basic Books, 2008), 47.

71 **Because programmers love acronyms:** Ensmenger, *The Computer Boys,* 96.

71 **Most programmers explicitly despise:** Ibid., 100–101.

71 **COBOL "cripples the mind":** To be fair, Dijkstra was a tough judge of programming languages. Of FORTRAN he wrote, "The sooner we can forget FORTRAN ever existed, the better, for as a vehicle of thought it is no longer accurate"; while PL/I, in his estimation "could turn out to be a fatal disease." Edsger W. Dijkstra, *Selected Writings on Computing: A Personal Perspective* (New York: Springer-Verlag, 1982), 130.

72 **COBOL: //Koh'Bol/, n:** *The New Hacker's Dictionary,* 3rd ed., comp. Eric S. Raymond (Cambridge, MA: MIT Press, 1996), 115.

72 **wrote this off as a "snob reaction":** Sammet, "COBOL Session," 266.

72 **"as much as any other single person":** Ibid., 4.

73 **Moser would also help out:** Denise Gürer, "Pioneering Women in Computer Science," *Communications of the ACM* 38(1): 45–54, https://courses.cs.washington.edu/courses/csep590/06au/readings/p175-gurer.pdf.

73 **machine made from code alone:** Abbate, *Recoding Gender,* 84.

74 **"both the expertise to devise solutions":** Abbate, *Recoding Gender,* 81.

CHAPTER FIVE: THE COMPUTER GIRLS

75 **"The Computer Girls," the magazine reported:** Lois Mandel, "The Computer Girls," *Cosmopolitan,* 1967, 52–56.

75 **"You have to plan ahead:** Ibid.

76 **"'Of course we like having the girls around'":** Nathan Ensmenger, "Making Programming Masculine," in *Gender Codes: Why Women Are Leaving Computing,* ed. Thomas Misa (Hoboken, NJ: Wiley, 2010).

76 **Some estimates peg female programmers:** Ibid.

76 **Many of these were dramatic:** Abbate, *Recoding Gender, 92.*

76 **"If one character, one pause":** Frederick Brooks, *The Mythical Man-Month: Essays on Software Engineering* (Boston: Addison-Wesley, 1975), 8.

77 **Others have cited a personality clash:** Ensmenger, *The Computer Boys,* 147.

77 **This wage discrimination:** Abbate, *Recoding Gender,* 90.

78 **The introduction of formal:** Ensmenger, *The Computer Boys,* 239.

78 **"began as women's work":** Ensmenger, "Making Programming Masculine."

78 **"brought with it unspoken ideas:** Abbate, *Recoding Gender,* 103.

78 **As she told a historian in 1968:** Hopper, interview with Merzbach, 1968, 17.

78 **"stereotypically feminine skills of communication":** Abbate, *Recoding Gender,* 109.

79 **In the 1990s, when I finally:** Plant, *Zeroes + Ones,* 33.

80 **"When computers were vast systems":** Ibid., 37.

CHAPTER SIX: THE LONGEST CAVE

83 **The first of these guides:** Later generations of enslaved guides would sell these fish to tourists to raise enough money to buy their own freedom.

84 **"a bowl of spaghetti":** Roger W. Brucker, "Mapping of Mammoth Cave: How Cartography Fueled Discoveries, with Emphasis on Max Kaemper's 1908

Map" (Mammoth Cave Research Symposia, Paper 4, October 9, 2008, http://digitalcommons.wku.edu/mc_reserch_symp/9th_Research_Symposium_2008/Day_one/4.)

84 **One nameless noodle:** Ibid.

85 **"like chocolate frosting":** Richard D. Lyons, "A Link Is Found Between Two Major Cave Systems," *New York Times,* December 2, 1972, www.nytimes.com/1972/12/02/archives/a-link-is-found-between-two-major-cave-systems-link-found-between-2.html?_r=0.

85 **"in the open truck bed":** Patricia P. Crowther, Cleveland F. Pinnix, Richard B. Zopf, Thomas A. Brucker, P. Gary Eller, Stephen G. Wells, and John P. Wilcox, *The Grand Kentucky Junction: A Memoir* (St. Louis, MO: Cave Books, 1984), 96.

85 **"It's an incredible feeling":** Roger W. Brucker and Richard A. Watson, *The Longest Cave* (Carbondale: Southern Illinois University Press, 1976), 213.

86 **Will ran a "map factory":** Ibid., 171.

86 **"plotting commands on huge rolls":** Dennis G. Jerz, "Somewhere Nearby Is Colossal Cave: Examining Will Crowther's Original 'Adventure' in Code and in Kentucky," *Digital Humanities Quarterly* (2007), www.digitalhumanities.org/dhq/vol/1/2/000009/000009.html.

86 **In 1969, BBN was contracted:** James Gillies and Robert Caillau, *How the Web Was Born: The Story of the World Wide Web* (Oxford: Oxford University Press, 2000), 15.

87 **A lifelong mountaineer:** Brucker and Watson, *The Longest Cave,* 171.

87 **"I get cold when he's not keeping me company":** Crowther et al., *The Grand Kentucky Junction,* 10.

87 **They stayed up late:** Ibid.,19–20.

87 **"*Now* I can sleep":** Ibid., 20.

88 **"The route is never in view":** Brucker and Watson, *The Longest Cave,* xvii.

88 ***no exploration without survey:*** Brucker, "Mapping of Mammoth Cave."

88 **"working rationally and systematically":** Joseph P. Freeman, *Cave Research Foundation Personnel Manual,* 2nd ed. (Cave City, KY: Cave Research Foundation, 1975).

88 **"pulled apart in various ways":** Julian Dibbell, "A Marketable Wonder: Spelunking the American Imagination," *Topic Magazine* 2, www.webdelsol.com/Topic/articles/02/dibbell.html.

88 **"software is the final victory":** Richard Powers, *Plowing the Dark* (New York: Picador, 2001), 307.

89 **"Another caver who was with":** Jerz, "Somewhere Nearby Is Colossal Cave."

90 **"harrowing of Hell":** Tracy Kidder, *The Soul of a New Machine* (New York: Back Bay Books, 1981), 88.

90 **By the time she encountered:** www.legacy.com/obituaries/dispatch/obituary.aspx?n=john-preston-wilcox&pid=145049233.

90 **"completely different from the real cave":** Jerz, "Somewhere Nearby Is Colossal Cave."

91 **"*Adventure*'s Colossal Cave, at least":** Walt Bilofsky, "Adventures in Computing," *Profiles: The Magazine for Kaypro Users* 2, no. 1 (1984): 25, https://archive.org/stream/PROFILES_Volume_2_Number_1_1984-07_Kaypro_Corp_US/PROFILES_Volume_2_Number_1_1984-07_Kaypro_Corp_US_djvu.txt.

91 **"the deep recesses you explored":** Steven Levy, *Hackers: Heroes of the Computer Revolution,* 25th Anniversary Edition (Sebastopol, CA: O'Reilly Media, 2010), 113.

92 **His daughters were told to use it:** Jerz, "Somewhere Nearby Is Colossal Cave."
92 **"analogous to the democratization of reading":** Mary Ann Buckles, "Interactive Fiction: The Computer Storygame 'Adventure'" (PhD thesis, University of California, San Diego, 1985).
92 **"a mythological urtext":** Espen J. Aarseth, *Cybertext: Perspectives on Ergodic Literature* (Baltimore, MD: Johns Hopkins University Press, 1997), 108.
93 **By 1984, the number of women:** Thomas J. Misa, "Gender Codes: Defining the Problem," in *Gender Codes: Why Women Are Leaving Computing,* ed. Thomas J. Misa (Hoboken: Wiley-IEEE Computer Society Press, 2010), 3.
93 **"If she can only cook":** "Bytes for Bites: The Kitchen Computer," Computer History Museum, www.computerhistory.org/revolution/minicomputers/11/362.
93 **"more authority, power, and intelligence":** Jesse Adams Stein, "Domesticity, Gender, and the 1977 Apple II Personal Computer," *Design and Culture* 3, no. 2 (2011): 193–216.

CHAPTER SEVEN: RESOURCE ONE

96 **"pueblo in the city":** Charles Raisch, "Pueblo in the City: Computer Freaks, Architects and Visionaries Turn a Vacant San Francisco Candy Factory into a Technological Commune," *Mother Jones,* May 1976, accessed February 5, 2017, https://books.google.com/books?id=aOYDAAAAMBAJ&lpg=PA27&dq=mother%20jones%20pueblo%20in%20the%20city%20charles%20raisch&pg=PA28#v=onepage&q&f=true.
97 **"big noodles" of industrial power:** Lee Felsenstein, interview with the author, March 7, 2017.
97 **"The school was on strike":** Pamela Hardt-English, interview with the author, February 6, 2017.
97 **"It was the first time many of them":** Optic Nerve, "Project One," 1972, Pacific Film Archive Film and Video Collection, https://archive.org/details/cbpf_000052.
98 **"Our vision was making technology accessible:** Hardt-English, interview with the author, February 6, 2017.
99 **"My brother came to live with me":** Ibid.
99 **"one of the great hustles of modern times":** Stewart Brand, "SPACEWAR: Fanatic Life and Symbolic Death Among the Computer Bums," *Rolling Stone,* December 7, 1972.
99 **"She had this way of sort of screwing up":** Felsenstein, interview with the author, March 7, 2017.
99 **"Pamela was about the only person":** Jane R. Speiser, *Roadmap of the Promised Land* (Turin: Edizioni Angolo Manzoni, 2006), 45.
100 **"If people needed something":** Hardt-English, interview with the author, February 6, 2017.
100 **"half or more of computer science is heads":** Stewart Brand, *II Cybernetic Frontiers* (New York: Random House, 1974), 49–50.
101 **"magnificent men with their flying machines":** Ibid., 50.
102 **totems of a "regimented order":** Lee Felsenstein, "Community Memory: The First Public-Access Social Media System," in *Social Media Archaeology and Poetics,* ed. Judy Malloy (Cambridge, MA: MIT Press, 2016), 91.
102 **"We opened the door to cyberspace":** Ibid., 89.
103 **as a "living metaphor":** Levy, *Hackers,* 128.

103 **"When I was at Project One":** Hardt-English, interview with the author, February 20, 2017.
104 **"Pam found herself unwittingly":** Felsenstein, e-mail message to the author, April 9, 2017.
104 **"The closest thing to a governing agency":** Sherry Reson, interview with the author, February 20, 2017.
104 **"Hippies with their old ladies":** Chris Macie, interview with the author, February 20, 2017.
105 **"We wanted the social workers":** Reson, interview with the author, February 20, 2017.
106 **"They figured out a way to put technology":** Joan Lefkowitz, interview with the author, March 6, 2017.
106 **"women didn't do the programming":** Ibid.
107 **"You're countering dominance behavior":** Reson, interview with the author, February 20, 2017.

CHAPTER EIGHT: NETWORKS

110 **"a scientist who needed":** Janet Abbate, *Inventing the Internet* (Cambridge, MA: MIT Press, 1999), 1.
111 **"looking kind of like unmade beds":** Elizabeth "Jake" Feinler, "Oral History of Elizabeth (Jake) Feinler: Interviewed by Marc Weber," September 10, 2009, Computer History Museum, 4, www.computerhistory.org/collections/oralhistories.
111 **While doing graduate work:** Elizabeth Jocelyn Feinler, "Interview by Janet Abbate," IEEE History Center, July 8, 2002, http://ethw.org/Oral-History:Elizabeth_%22Jake%22_Feinler.
111 **Realizing she was more interested:** Ibid.
112 **"Mother of all Demos":** It also ran on an SDS-940—and according to some accounts, the very same machine that eventually made its way to Resource One.
112 **"He would come down and say":** Elizabeth "Jake" Feinler, interview with the author, September 1, 2017.
112 **The connection crashed halfway through:** Leonard Kleinrock, "An Early History of the Internet [History of Communications]," *IEEE Communications Magazine* 48, no. 8 (August 2010).
112 **"I said, 'What's a Resource Handbook?'":** Feinler, interview with the author, September 1, 2017.
112 **"It was pretty obvious":** Ibid.
113 **"the kids ran the machine":** Ibid.
113 **Despite these challenges, the Resource Handbook:** Garth O. Bruen, *WHOIS Running the Internet: Protocol, Policy, and Privacy* (Hoboken, NJ: Wiley, 2016), 27.
114 **Ellen Westheimer, who worked at Bolt:** Elizabeth Feinler, "Host Tables, Top-Level Domain Names, and the Origin of Dot Com," *IEEE Annals of the History of Computing* 33, no. 3 (March 2011), http://ieeexplore.ieee.org/stamp/stamp.jsp?arnumber=5986499.
115 **"There weren't many women who were programmers":** Feinler, interview with the author, September 1, 2017.
115 **"somebody came and yelled at me":** Ibid.

115 **"It was harder to get higher-ups":** Elizabeth "Jake" Feinler, interview by Marc Weber, 14.
115 **"If you didn't know where else to go":** Ibid.
116 **"It was just unending":** Feinler, interview with the author, September 1, 2017.
116 **"We had a woman come in and clean":** Ibid.
116 **At a certain point, she began to remove her own name:** Ibid.
116 **"That was almost from the beginning":** Ibid.
117 **"We were just trying to build things":** Ibid.
117 **"There was a couch in there":** Feinler, interview by Weber, 9.
118 **"I always meant to get married":** Feinler, interview with the author, September 1, 2017.
118 **"The Internet was more fun than a barrel of monkeys":** Internet Society, "Elizabeth Feinler—INTERNET HALL OF FAME PIONEER," 4:40, filmed April 23, 2012, posted to YouTube May 8, 2012, https://youtu.be/idb-7Z3qk_o.
118 **"She wanted to have a crab feast":** Mary K. Stahl, interview with the author, September 7, 2017.
118 **"It was like my family":** Feinler, interview with the author, September 1, 2017.
119 **"a kid hacker would be talking":** Feinler, interview by Weber, 19.
119 **"WHOIS was probably one of our biggest servers":** Ibid., 26.
120 **WHOIS does nothing less:** Bruen, *WHOIS Running the Internet,* 7.
120 **"a burn-out job":** Stahl, interview with the author, September 7, 2017.
120 **"cumbersome and inefficient":** Feinler, "Host Tables."
120 **Jake suggested dividing them into generic categories:** Ibid.
122 **"I think there was a lot of bad feeling":** Stahl, interview with the author, September 7, 2017.
122 **"The main purpose of the Internet":** Feinler, interview with the author, September 1, 2017.
122 **"coming up with the best suite of protocols":** Ibid.
123 **"Gee, that person looks out of place":** Rebecca J. Rosen, "Radia Perlman: Don't Call Me the Mother of the Internet," *The Atlantic,* March 3, 2014, www.theatlantic.com/technology/archive/2014/03/radia-perlman-dont-call-me-the-mother-of-the-internet/284146.
123 **"She was busy freeing the johns":** Feinler, interview by Marc Weber, 28.
124 **"The more senior you get":** Radia Perlman, interview with the author, June 22, 2017.
124 **"I never took anything apart":** Ibid.
124 **"Certainly, people like that":** Ibid.
125 **"He said, 'Are you happy professionally?'":** Ibid.
125 **"I always had the world's worst cold":** Ibid.
125 **"by issuing a memo saying":** Ibid.
126 **"stepping outside of the complexity":** Ibid.
126 **"I try to design things":** Ibid.
126 **"invent a magic box":** Imagining the Internet, "Internet Hall of Fame 2014: Radia Perlman," 16:48, filmed April 7, 2014, posted to YouTube April 15, 2014, https://youtu.be/G3zJuMht5Kk.
126 **"He thought that was going to be hard":** Perlman, interview with the author, June 22, 2017.
126 **"I realized, oh wow—it's trivial":** Ibid.

127 **"Without me, if you just blew":** Imagining the Internet, "Internet Hall of Fame 2014: Radia Perlman," 16:48, filmed April 7, 2014, posted to YouTube April 15, 2014, https://youtu.be/G3zJuMht5Kk.

127 **"She said, 'Certainly,' and she quoted":** Perlman, interview with the author, June 22, 2017.

127 **From this, Radia adapted her spanning-tree algorithm:** Radia Perlman, *Interconnections: Bridges, Routers, Switches, and Internetworking Protocols* (Boston: Addison-Wesley, 2000), 58.

128 **"If I do my job right":** Perlman, interview with the author, June 22, 2017.

CHAPTER NINE: COMMUNITIES

129 **"I learned to drive":** Reyner Banham, *The Architecture of Four Ecologies* (Berkeley: University of California Press, 2009), 5.

130 **"rather like taking a tank":** Abbate, *Inventing the Internet*, 2.

130 **Rather, this next generation:** That convenience was no accident. The two men who wrote the original BBS software in 1978 did so because their Chicago microcomputer club wanted to share newsletters, even during harsh midwestern snowstorms.

131 **"touched her core":** Madeline Gonzales Allen, "Community Networking, an Evolution," in *Social Media Archaeology and Poetics*, 291.

131 **In a five-part documentary:** Jason Scott, *BBS: The Documentary*, https://archive.org/details/BBS.The.Documentary.

131 **"Van Halen rules":** Ibid.

132 **"first realization that there were people":** "First Memories: Aliza Sherman," Women's Internet History Project, last modified March 26, 2015, http://womensinternethistory.org/2015/03/first-memories-aliza-sherman.

132 **"All software does is manage symbols:** Stewart Brand, *Whole Earth Software Catalog* (New York: Quantum Press/Doubleday, 1984), 4.

132 **"wanted to have an experiment":** Nancy Rhine, interview with the author, February 8, 2017.

133 **"it was like somebody had literally":** Naomi Pearce, interview with the author, February 16, 2017.

133 **Or, as The WELLbeings say:** Cliff Figalo, "The WELL: A Regionally Based On-Line Community on the Internet," in *Public Access to the Internet*, eds. Brian Kahin and James Keller (Cambridge, MA: MIT Press, 1995), 55.

133 **"The WELL stood for Whole Earth 'Lectronic Link":** Rhine, interview with the author, February 8, 2017.

134 **journalists, ex-hippies, and hobbyist:** *Encyclopedia of New Media*, ed. Steve Jones (Thousand Oaks, CA: Sage Reference Publications, 2003), 481.

134 **"I can't send beams":** Horn, *Cyberville*, 72.

135 **"It wasn't like I was a visionary":** Stacy Horn, interview with the author, May 26, 2016.

135 **"people just *openly* made fun":** Ibid.

135 **"to do something which sounded":** Ibid.

136 ***Go away,* they'd scream at the phone:** Horn, *Cyberville*, 44.

136 **"like Cracker Jacks":** Horn, interview with the author, May 26, 2016.

136 **"deserted, empty, time tripping":** Horn, *Cyberville*, 44.

136 **"Clinton and Gore were":** Horn, interview with the author, May 26, 2016.

137 **"people started having this sense":** Ibid.

137 **"When computer people came online":** Casey Kait and Stephen Weiss, *Digital Hustlers: Living Large and Falling Hard in Silicon Alley* (New York: HarperCollins, 2001), 56.

137 **"There was a contingent of people":** Marisa Bowe, interview with the author, July 26, 2016.

137 **"punk rock suburbanite-city girl":** Horn, *Cyberville,* 147.

138 **"Sometimes newcomers don't realize":** Marisa Bowe, "Net Living: The East Coast Hang Out," *Wired,* March 1, 1993, www.wired.com/1993/03/net-living-the-east-coast-hang-out.

138 **"I hate myself for being a fucking addict":** Horn, *Cyberville,* 76.

139 **"My success was due in part":** Stacy Horn, "Echo," in *Social Media Archaeology and Poetics,* 246.

139 **"We would have these meetings":** Horn, interview with the author, May 26, 2016.

140 **"She was a master at it":** Ibid.

140 **"This is Stacy Horn":** Aliza Sherman, interview with the author, June 2, 2016.

140 **"In those days," Stacy writes:** Horn, *Cyberville,* 92.

141 **"I talk differently when I'm with my choir friends":** Horn, interview with the author, May 26, 2016.

142 **"WIT with a leather jacket":** Horn, *Cyberville,* 246.

142 **"Shades of separate but equal":** Ibid., 87.

143 **"Back then it was just so odd":** "First Memories: Aliza Sherman."

143 **a long list of gender options:** The popular MUD LambdaMOO offered eleven genders. A character could be male or female, but it could also be plural, appearing as a kind of colony, or ego, for which the only pronoun is "I," or else royal, using only the royal "we." The neutral genders had their own pronoun conventions: "splat" relied on asterisks, while a "spivak" character was referred to as "e," "em," "eir," "emself."

143 **"interested in seeing how the other half lives":** Pavel Curtis, "Mudding: Social Phenomena in Text-Based Virtual Realities," https://w2.eff.org/Net_culture/MOO_MUD_IRC/curtis_mudding.article.

143 **"On the nets":** Allucquére Rosanne Stone, "Will the Real Body Please Stand Up?: Boundary Stories About Virtual Cultures," in *Cyberspace: First Steps,* ed. Michael Benedikt (Cambridge, MA: MIT Press, 1992), 84.

144 **"I didn't know what to do":** Horn, interview with the author, May 26, 2016.

144 **"the George Wallace of cyberspace":** Horn, *Cyberville,* 87.

144 **"Cyberspace makes it easier":** Ibid., 102.

145 **"All right, she begins":** Stacy Horn, e-mail message to the author, February 26, 2016.

145 **"There were *laundromats, delis*":** Horn, interview with the author, May 26, 2016.

146 **"It was this weird thing":** Ibid.

146 **83 percent of Echoids:** Echo NYC About Page, December 1998, https://web.archive.org/web/19990508065020/http://www.echonyc.com/about.

146 **"the bedrock of Silicon Alley":** Jason Cherkovas, "New York's New Media Ground Zero," in *Silicon Alley: The Rise and Fall of a New Media District,* ed. Michael Indergaard (New York: Routledge, 2004), 32.

146 **A group of Echoids got together:** Howard Mittelmark, interview with the author, July 21, 2016.
147 **"The strongest virtual communities":** Horn, *Cyberville,* 113.
147 **"Stacy was more of an autocrat":** Mittelmark, interview with the author, July 21, 2016.
148 **"he was, like, really subversive":** Bowe, interview with the author, July 26, 2016.
148 **"When the world that you're in":** Ibid.
148 **"Echo is Echo because of the hosts":** Horn, *Cyberville,* 39.
148 **"*animateurs*" culled from:** Howard Rheingold, *The Virtual Community: Homesteading on the Electronic Frontier* (Cambridge, MA: MIT Press, 2000), 235.
149 **"Hosts are the people":** Rheingold, *The Virtual Community,* 26.
149 **"front page of the Internet":** Adrian Chen, "The Laborers Who Keep Dick Pics and Beheadings Out of Your Facebook Feed," *Wired,* October 23, 2014, www.wired.com/2014/10/content-moderation.
149 **"cyberaffirmative action":** Horn, *Cyberville,* 96.
149 **"I heard women talking":** Mittelmark, interview with the author, July 21, 2016.
150 **"A PLANE JUST CRASHED":** Horn, "Echo," 245.
150 **"The hottest topic":** Horn, *Cyberville,* 53.
150 **"Someone in the twenty-second century":** Horn, interview with the author, May 26, 2016.

CHAPTER TEN: HYPERTEXT

154 **Sprawling, self-referential novels:** Jay David Bolter, *Writing Space: The Computer, Hypertext, and the History of Writing* (Hillsdale, NJ: Lawrence Erlbaum Associates, 1991), 24.
154 **"This web of time":** Jorge Luis Borges, *Ficciones* (New York: Grove Press, 1962), 100.
155 **"Every cow, every sheep":** Dame Wendy Hall, interview with the author, January 18, 2017.
156 **"like those of the Last Judgement":** Richard FitzNeal (Richard Fitz Nigel), *Dialogus de Scaccario, the Course of the Exchequer,* and *Constitutio Domus Regis, the King's Household,* ed. and trans. Charles Johnson (Oxford: Oxford University Press, 1983), 64.
156 **"that's the British Broadcasting Corporation":** Hall, interview with the author, January 18, 2017.
156 **Robot limbs will be used:** "1986: A Child's View of the Future," *Domesday Reloaded,* BBC, www.bbc.co.uk/history/domesday/dblock/GB-424000-534000/page/16.
156 **"The ideas were stunning":** Hall, interview with the author, January 18, 2017.
157 **"shy and retiring student":** Web Science Trust, "Professor Wendy Hall: Making Links," 50:28, filmed July 14, 1997, posted to YouTube March 12, 2017, https://youtu.be/cFa3e-VkgMk.
157 **"I was happy in my world":** Wendy Hall, interviewed by Jim Al-Khalili, October 8, 2013, *The Life Scientific,* BBC Radio 4.
157 **"I began to see the future":** Hall, interview with the author, January 18, 2017.
157 **"One professor told me in public":** Ibid.
158 **They bombed the boat:** "1979: IRA Bomb Kills Lord Mountbatten," *On This Day August 27,* BBC, http://news.bbc.co.uk/onthisday/hi/dates/stories/august/27/newsid_2511000/2511545.stm.

158 **"Their spare time, of which they have a lot":** "1986: Punks in Romsey," *Domesday Reloaded,* BBC, www.bbc.co.uk/history/domesday/dblock/GB-432000-120000/page/4.

159 **the library inherited some:** Web Science Trust, "Professor Wendy Hall: Making Links," 50:28, filmed July 14, 1997, posted to YouTube March 12, 2017, https://youtu.be/cFa3e-VkgMk.

159 **"The archivist came to see me":** Hall, interview with the author, January 18, 2017.

161 **"Links in themselves are a valuable store":** W. Hall and D. Simmons, "An Open Model for Hypermedia and Its Application to Geographical Information Systems," Proceedings of Eurographics '92, Cambridge, UK.

162 **"the whole foundation of hypertext":** Nicole Yankelovich, interview with author, January 9, 2017.

163 **"If I write something that doesn't work":** Cathy Marshall, interview with the author, January 11, 2017.

163 **the summer she matriculated:** Cathy Marshall, "The Freshman: Confessions of a CalTech Beaver," in *No Middle Initial,* February 25, 2011, http://ccmarshall.blogspot.com.

163 **"my housewifing skills":** Cathy Marshall, interview with the author, December 19, 2016.

163 **whorehouse keeping down the "fucking overhead":** Ibid.

163 **corduroy beanbag chairs:** One, a corduroy number in an unmistakably seventies ocher, is on display at the Computer History Museum, alongside a video of the software engineer Adele Goldberg talking about how difficult it was to sit on the beanbag chairs while pregnant. "Once you sunk in," she explains, "you couldn't jump up." Which is to say, there weren't very many women at Xerox PARC. But those who were there did exceptional things, www.computerhistory.org/revolution/input-output/14/348/2300.

164 **"The thing I loved most about PARC":** Marshall, interview with the author, December 19, 2016.

164 **Her partner, Judy Malloy, a poet:** Cathy Marshall and Judy Malloy, "Closure Was Never a Goal of This Piece," in *Wired Women: Gender and New Realities in Cyberspace,* ed. Lynn Cherny and Elizabeth Reba Weise (Seattle: Seal Press, 1996), 64–65.

164 **"the way you wrote papers":** Marshall, interview with the author, January 11, 2017.

164 **Hypertext is to text:** Rob Swigart, "A Writer's Desktop," in *The Art of Human-Computer Interface Design,* eds. Brenda Laurel (Boston: Addison-Wesley, 1990), 140.

165 **NoteCards became a vital tool:** Randall H. Trigg and Peggy M. Irish, "Hypertext Habitats: Experiences of Writers in NoteCards," HYPERTEXT '87, Proceedings of the ACM Conference on Hypertext, 89–108.

165 **In 1987, Apple released:** Apple's idea of digital notecards was strikingly similar to what was being developed at Xerox PARC, but it might have been in the water, too—borrowing a familiar office metaphor was "one of those ideas that could be pervasive," Cathy says, democratically.

165 **"rueful sense that this was":** Esther Dyson, *Release 1.0,* November 25, 1987.

165 **"The hypertext conferences were lovely":** Hall, interview with the author, January 18, 2017.

165 **"Computer science has always marginalized":** Marshall, interview with the author, January 11, 2017.

165 **"There were little islands":** Ibid.

166 **NoteCards allowed multiple arrangements:** Trigg and Irish, "Hypertext Habitats," 89–108.

166 **"difficult to articulate within the bounds":** Catherine C. Marshall, Frank M. Shipman III, James H. Coombs, "VIKI: Spatial Hypertext Supporting Emergent Structure," ECHT '94, Proceedings of the 1994 ACM European Conference on Hypermedia technology, 13–23.

167 **"spatial holding patterns":** Alison Kidd, "The Marks Are on the Knowledge Worker," CHI '94, Proceedings of the SIGCHI Conference on Human Factors in Computing Systems, 186–91.

167 **The conference floor, a hotel reception:** "List of Demonstrators: Hypertext '91," World Wide Web Consortium, www.w3.org/Conferences/HT91/Denoers .html.

168 **ten-thousand-dollar jet-black NeXT cube:** Interestingly, a female student on a one-year work placement at CERN, Nicola Pellow, had written a no-frills, text-based Web browser that could run on any computer—the Line Mode browser—but it was utilitarian, and none too flashy. Berners-Lee was willing to lug the NeXT across the globe in order to present the concept of a graphical Web site.

168 **"He said you needed an Internet connection":** Marshall, interview with the author, December 19, 2016.

168 **"I was looking at it":** Hall, interview with the author, January 18, 2017.

169 **"That was all considered counter":** Marshall, interview with the author, December 19, 2016.

169 **"hypertext-like interface":** Lynda Hardman, "Hypertext '91 Trip Report," *ACM SIGCHI Bulletin* 24, no. 3 (July 1, 1992).

169 **Cathy Marshall proposed that:** Frank M. Shipman III, Catherine C. Marshall, and Mark LeMere, "Beyond Location: Hypertext Workspaces and Non-Linear Views," in Proceedings of ACM Hypertext '99, Darmstadt, Germany, 121–30.

170 **"I'm not sure exactly how to describe it":** Wendy Hall, "Back to the Future with Hypertext: A Tale of Two or Three Conferences," in Proceedings of ACM 18th Conference on Hypertext and Hypermedia 2007, Manchester, UK, 179–180.

170 **according to a 2013 study:** Jason Hennessey and Steven Xijin, "A cross disciplinary study of link decay and the effectiveness of mitigation techniques," *BMC Bioinformatics201314* (Suppl. 14): S5. http://bmcbioinformatics.biomedcentral .com/articles/10.1186/1471-2105-14-S14-S5.

171 **"Claire's writing a book about me":** Dame Wendy Hall, interview with the author, March 1, 2017.

172 **"through a Microcosm viewer":** Ibid.

172 **didn't suffer from dead links:** Les Carr, Wendy Hall, Hugh Davis, and Rupert Hollom, "The Microcosm Link Service and Its Application to the World Wide Web," in Proceedings of the First World-Wide Web Conference, 1994, Geneva.

172 **"links to many different destinations":** Hall, interview with the author, March 1, 2017.

173 **"By enhancing the Web":** Carr et al., "The Microcosm Link Service."

173 **"People used to say":** Hall, interview with the author, March 1, 2017.

173 **"The Web has shown us":** Web Science Trust, "Professor Wendy Hall: Making Links."

174 **"We are now, twenty-seven years after":** Hall, interview with the author, February 28, 2017.

174 **"That was the core of Microcosm":** Ibid.

CHAPTER ELEVEN: MISS OUTER BORO

177 **"Communications media often seem":** Abbate, *Inventing the Internet,* 5.

178 **"I didn't have any interest in programming":** Bowe, interview with the author, July 26, 2016.

179 **"science fiction, women's rights:** "Between PLATO and the Social Media Revolution," May 10, 1983, http://just.thinkofit.com/between-plato-and-the-social-media-revolution.

179 **including men impersonating women:** David R. Woolley, "PLATO: The Emergence of Online Community," in *Social Media Archaeology and Poetics,* ed. Judy Malloy (Cambridge, MA: MIT Press, 2016), 115.

179 **Trying LSD for the first time:** Marisa Bowe, "Wednesday September 18, 1974," Staff Homepages, Word.com, http://web.archive.org/web/19970615070535/http://www.word.com:80/newstaff/mbowe/one/date30b.html.

179 **"Donald and Ivana Trump":** Marisa Bowe, "When I Grow Up," Vice.com, November 30, 2004, www.vice.com/read/when-I-v11n3.

179 **"This is just like PLATO":** Bowe, interview with the author, July 26, 2016.

180 **"I'm *allergic* to the Grateful Dead":** Digital Archaeology, "www.word.com, circa 1995," 5:18, posted to YouTube June 2011, www.youtube.com/watch?v=mxEhqmpymnQ.

180 **"I logged on to Echo":** Bowe, interview with the author, July 26, 2016.

180 **"the idea that you could converse":** Ibid.

180 **"They felt sorry for me":** Ibid.

180 **"She makes me spit out my coffee":** Horn, *Cyberville,* 21.

181 **"She knows she's smart":** Ibid., 78.

181 **"Henry James of the Alley":** Kait and Weiss, *Digital Hustlers,* 78.

181 **"It was like a mini-celebrity":** Bowe, interview with the author, July 26, 2016.

182 **"The video art scene was overrun":** Jaime Levy, interview with the author, August 6, 2016.

182 **"on a Citibank scholarship":** Jaime Levy, "Web Content Producer," in *Gig: Americans Talk About Their Jobs,* eds. John Bowe, Marisa Bowe, and Sabin Streeter (New York: Three Rivers Press, 2000), 364.

182 **"I think she saw the opportunity":** Levy, interview with the author, August 6, 2016.

183 **"if you hate it, take the files off":** Jaime Levy, "Jaime Levy and Electronic Publishing from Life and Times KCET in 1993," 5:33, posted to YouTube June 2015, https://youtu.be/t5aQCQ7-WYU.

183 **"We called it Smell-o-vision":** Ibid.

184 **"my, sorta, digital graffiti":** Ibid.

184 **Jaime's disks, packaged on floppy:** Jaime Levy, *UX Strategy: How to Devise Innovative Digital Products That People Want* (Sebastopol, CA: O'Reilly Media, 2015), 119–22.

184 **six thousand copies at six bucks a pop:** Jaime Levy, "*Dateline NBC*—'Can You Be a Millionaire? featuring Jaime Levy (2000),'" 10:19, posted to YouTube September 2016, https://youtu.be/v__oMcjFkI0.

184 **"I was the Kurt Cobain":** Austin Bunn, "Upstart Start-ups," *Village Voice,* November 11, 1997, www.villagevoice.com/news/upstart-start-ups-6423803.
184 **"Jaime really knew how to present herself":** Kait and Weiss, *Digital Hustlers,* 79.
185 **She was a hacker through:** "Designer Dossier: Jaime Levy, Cyberslacker," *Computer Player,* June 1994, www.ehollywood.net/presskit/computerplayer/body.htm.
185 **"If you have never seen":** Levy, "Jaime Levy and Electronic Publishing from Life and Times KCET in 1993."
186 **"He just really wanted to party":** Levy, interview with the author, Los Angeles, August 6, 2016.
186 **"bonehead interface design":** "Designer Dossier: Jaime Levy, Cyberslacker."
186 **she was often mistaken for a janitor:** "IBM'S Cyberslacker," *New York,* June 13, 1994, http://jaimelevy.com/press/newyork2.htm.
186 **"Who's going to buy this shit?":** Levy, interview with the author, August 6, 2016.
186 **"Once the browser came out":** Ibid.
186 **She quit her day job:** Andrew Smith, *Totally Wired: The Wild Rise and Crazy Fall of the First Dotcom Dream* (New York: Simon & Schuster, 2012).
186 **Multimillion-dollar startups:** Kait and Weiss, *Digital Hustlers,* 225–27.
187 **Over two adjoining warehouses:** *We Live in Public.*
187 **"Josh Harris always *says* he didn't copy me":** Levy, interview with the author, August 6, 2016.
187 **"principled slackers, arty punk rockers":** Michael Indergaard, *Silicon Alley: The Rise and Fall of a New Media District* (New York: Routledge, 2004), 1.
187 **"That's all that was up there":** Levy, interview with the author, August 6, 2016.
187 **"Nineteen ninety-five is cool":** Vanessa Grigoriadis, "Silicon Alley 10003," *New York,* March 6, 2000, http://nymag.com/nymetro/news/media/Internet/2285.
188 **"It turned us all into apostles":** Kait and Weiss, *Digital Hustlers,* 47.
188 **"I didn't know what she meant":** Bowe, interview with the author, July 26, 2016.
188 **"pretend like I'm some super-straight":** Ibid.
189 **Icon thought they'd make millions:** Kait and Weiss, *Digital Hustlers,* 78.
189 **"knew how to approach a medium":** Bowe, interview with the author, July 26, 2016.
189 **"hooked on amateur stuff":** Marisa Bowe, in "Wiring the Fourth Estate: Part One of the FEED Dialog on Web Journalism," 1996, https://web.archive.org/web/19970225063402/http://www.feedmag.com/96.06dialog/96.06dialog1.html.
189 **Marisa went so far as to publish:** The day after her prom: "I can't stand the word *diary,* the word reminds me of diarrhea. And of a little pink-frocked, rosy-cheeked, curly-haired, moony-eyed chick writing a daily account of her sweet little heart's ups & downs."
189 **Her very first editorial was an abridged life story:** Marisa Bowe, "Letter from the Editor," Word.com, Autumn 1995, http://web.archive.org/web/19990912085004/http://www.word.com/info/letter/index.html.
190 **"What was fundamentally most fascinating":** Bowe, interview with the author, July 26, 2016.
190 **"In the world of *Word*":** "The Thirty Most Powerful Twentysomethings in America: Jaime Levy," *Swing* magazine, January 1996, http://jaimelevy.com/press/swing.htm.

190 **"Even the name 'Webmate'":** Bowe, interview with the author, July 26, 2016.

191 ***Word* was logging ninety-five thousand:** Steve Silberman, "Word Down: The End of an Era," *Wired,* March 11, 1998, https://web.archive.org/web/20080425015225/http://www.wired.com/culture/lifestyle/news/1998/03/10829.

191 ***Newsweek* announced *Word*:** "So What's a Web Browser, Anyway?" *New York Times,* August 14, 1995, http://jaimelevy.com/press/newyorktimes.htm.

191 **"Beavis and Butt-head of the Internet":** Michael Kaplan, "Word Up?!," *Digital Creativity,* June–July 1996, 37.

191 **"more like a rock band":** Kait and Weiss, *Digital Hustlers,* 79.

191 **"She would literally *hiss* about text":** Marisa Bowe, interview with the author, August 29, 2016.

191 **"people who were getting filthy rich":** Bowe, interview with the author, August 29, 2016.

192 **"All the attention was on me":** Levy, interview with the author, August 6, 2016.

192 **"Jaime Levy has the last Word":** Jason Calcanis, "Cybersurfer's Silicon Alley," *PAPER,* March 1996, 122.

192 **"I've never been a person to last":** Ibid.

192 **"To pass that up"** Bowe, interview with the author, August 29, 2016.

192 **"the death of the Web as we knew it":** Indergaard, *Silicon Alley,* 1.

193 **"was one of these Japanese kids":** Bowe, interview with the author, July 26, 2016.

193 **"He's an artist":** Ibid.

193 **"We were essentially artists and bohemians":** Ibid.

193 **sold ads at around $12,500 a pop:** Kaplan, "Word Up?!"

193 **"For us, the creative team":** Naomi Clark, interview with the author, March 25, 2017.

193 **"Going public was the business model":** Bowe, interview with the author, July 26, 2016

194 **"D. B. Cooper–ish disappearance":** "Hit & Run 05.31.01," www.suck.com/daily/2001/05/31.

194 ***Wired* called it the end of an era:** Silberman, "Word Down."

194 **"It made so little sense":** Clark, interview with the author, March 25, 2017.

194 **Their new owner, Zapata Corporation:** Lisa Napoli, "From Oil to Fish to the Internet: Zapata Tries Another Incarnation," *New York Times,* May 18. 1998.

194 **took out full-page ads:** Kaitlin Quistgaard, "On the Edge and Under the Wing," *Wired,* September 1, 1998, https://web.archive.org/web/20101107173917/http://www.wired.com/culture/lifestyle/news/1998/09/14682.

194 **"subgeniuses who smoked too much pot":** Silberman, "Word Down."

195 **"I was like, okay, why don't we":** Levy, interview with the author, August 6, 2016.

195 **"production studio for the Internet":** Levy, "Web Content Producer," 364.

196 **The office was a fake-out:** Bumming around Electronic Hollywood in those days were *Psychic TV*'s Genesis P-Orridge, who makes a cameo in the *Dateline* piece, Clay Shirky, and "Tanya," whose last name Jaime can't remember but is pretty sure was in the virtual dildo business.

196 **Jaime expressed this ongoing hustle in a rap:** Grigoriadis, "Silicon Alley 10003."

197 **"Welcome to MTV's online":** Jaime Levy, "CyberSlacker–Episode 7 (Job Hunting Blues)," 5:41, posted to YouTube May 11, 2012, https://youtu.be/9rjVSRssz04.

197 **"It looks like a jar of salsa":** Jaime Levy, "CyberSlacker, Episode 8 (The Secret Sauce)," 6:05, posted to YouTube May 11, 2012, https://youtu.be/DbB0xBX9yEE
198 **it went public in 1999:** Indergaard, *Silicon Alley,* 149.
198 **not being cool enough:** Ibid., 139.
198 **"We pissed away almost all the money":** Levy, "Web Content Producer," 367.
198 **"Something was coming to a head":** Levy, interview with the author, August 6, 2016.
198 **Internet companies running out of money:** Indergaard, *Silicon Alley,* 142.
198 **the NASDAQ dropped below:** Kait and Weiss, *Digital Hustlers,* 297.
199 **"Within two months":** Levy, interview with the author, August 6, 2016.
199 **"We ate from the trough":** Charlie Leduff, "Dot-Com Fever Followed by Bout of Dot-Com Chill; What a Long, Strange Trip: Pseudo.Com to Dot.Nowhere," *New York Times,* October 27, 2000, www.nytimes.com/2000/10/27/nyregion/dot-com-fever-followed-bout-dot-com-chill-what-long-strange-trip-pseudocom.html?_r=0.
199 **Razorfish ousted its founders:** Indergaard, *Silicon Alley,* 150.
199 **"tombs built for Chinese emperors":** Ibid., 154.
199 **"our game plan for the next six months":** Levy, interview with the author, August 6, 2016.
200 **"Normally someone like me":** Bowe, interview with the author, August 29, 2016.
200 **"That's bullshit," Jaime says:** Levy, interview with the author, August 6, 2016.
200 **"Smallest violin for the fucking cyberkids":** Ibid.
200 **"coming after the collapse":** Indergaard, *Silicon Alley,* 160.
201 **by 2003, unemployment in New York:** Ibid., 159.
201 **"fucked up and left it on the subway":** Levy, interview with the author, August 6, 2016.
201 **"I started thinking of it as being":** Bowe, interview with the author, July 26, 2016.
201 **"We figured that in the end":** "Hit & Run 05.31.01."

CHAPTER TWELVE: WOMEN.COM

205 **"wanting to get an online community started":** Rhine, interview with the author, February 8, 2017.
206 **"access information and resources instantly":** Women's WIRE, *Women's Information Resource & Exchange* brochure (c. 1993).
206 **The headline was actually more unbelievable:** Mike Langberg, "Women Aim to Build an On-line World That Excludes Boors, Cybermashers," *San Jose Mercury News,* October 1, 1993.
206 **"it was much more of a positive":** Ellen Pack, interview with the author, March 3, 2017.
207 **The service hosted domestic abuse resources:** Connie Koenenn, "Chatting the High-Tech Way, on the Women's Wire," *Los Angeles Times,* February 24, 1994.
207 **"phone assault on the White House":** Miriam Weisang Misrach, "Only Connect," *Elle,* February 1994.
208 **"foster diversity . . . people with":** Leslie Regan Shade, "Gender and the Commodification of Community," in *Community in the Digital Age: Philosophy and*

Practice, eds. Andrew Feenberg and Darin Barney (Lanham, MD: Rowman & Littlefield Publishers, 2004), 145.

208 **"I liked the community piece":** Pack, interview with the author, March 3, 2017.

208 **"We had this big raging debate":** Pearce, interview with the author, February 16, 2017.

208 **"Where is the quilting bee":** Nancy Rhine, interview with the author, February 8, 2017.

209 **"had a lot of alpha males posturing":** Rhine, interview with the author, February 21, 2017.

209 **"were so polite and nice":** Rhine, interview with the author, February 8, 2017.

210 **"The original model had been":** Rhine, interview with the author, February 21, 2017.

210 **"Facebook is good because it creates community":** Max Read, "Does Even Mark Zuckerberg Know What Facebook Is?" *New York,* October 1, 2017, http://nymag.com/selectall/2017/10/does-even-mark-zuckerberg-know-what-facebook-is.html.

210 **the year the Web broke:** Myra M. Hart and Sarah Thorp, "Women.com," *HBS 9-800-216* (Boston: Harvard Business School Publishing, 2000), 4.

210 **"I always wanted to make this":** Pack, interview with the author, March 3, 2017.

210 **"like working on a rocket ship":** Marleen McDaniel, interview with the author, March 9, 2017.

210 **Ellen sent her a business plan:** Hart and Thorp, "Women.com," 5.

211 **"Some got it, of course":** Pack, interview with the author, March 3, 2017.

211 **"It was not easy to raise money":** McDaniel, interview with the author, March 9, 2017.

211 **"When I finally got a smaller":** Ibid.

211 **"That was a defining moment":** Ibid.

211 **The first big deal Marleen:** "Women's Wire Retools Its Goals," *Examiner Staff Report,* August 31, 1995, SFGate.com, www.sfgate.com/business/article/Women-s-Wire-retools-its-goals-3132265.php.

212 **"They bought our subscribers":** McDaniel, interview with the author, March 9, 2017.

212 **"They're saying it's a partnership":** David Plotnikoff, "Women's Wire Gives Up Ghost on Halloween," *Salt Lake Tribune,* October 30, 1995.

212 **"She wanted to run the company":** Ibid.

212 **"ranging from Barbie to Bosnia":** Janet Rae Dupree, "Women's Wire: Bosnia to Barbie," *York Daily Record,* August 12 1996.

212 **"The model switched to eyeballs":** Rhine, interview with the author, February 21, 2017.

213 **"I said, 'Oh my God'":** Gina Garrubbo, interview with the author, March 24, 2017.

213 **"Marleen, Ellen, me, and our CFO":** Ibid.

214 **"it's a woman's World Wide Web:** Anne Rickert and Anya Sacharow, "It's a Woman's World Wide Web," Media Metrix and Jupiter Communications, August 2000.

214 **"When we jumped off the cliff":** McDaniel, interview with the author, March 9, 2017.

214 **Because women controlled more than 80 percent:** Hart and Thorp, "Women.com," 4.

214 **While the original Women's WIRE:** Janet Rae-Dupree, "Women's Wire Blends Humor, Off-Beat Info Online," *San Jose Mercury News,* August 5, 1996.

214 **by 1996, women.com was getting:** Shade, "Gender and the Commodification of Community," 145.

215 **"We were on the map":** Pack, interview with the author, March 3, 2017.

215 **According to an industry research:** Hart and Thorp, "Women.com," 4.

215 **"We built the revenue":** Gina Garrubbo, interview with the author, March 24, 2017.

215 **"emphasized current news":** Shade, "Gender and the Commodification of Community," 145.

216 **ChickClick, which began as a small:** Janelle Brown, "What Happened to the Women's Web?" Salon.com, August 25, 2000, www.salon.com/2000/08/25/womens_web.

216 **"She was scandalously interesting":** McDaniel, interview with the author, March 9, 2017.

216 **iVillage was expert at capitalizing:** Erik Larson, "Free Money: The Internet IPO That Made Two Women Rich, and a Lot of People Furious," *New Yorker,* October 11, 1999.

217 **"like two cars on a racetrack":** McDaniel, interview with the author, March 9, 2017.

217 **"We competed for everything":** Garrubbo, interview with the author, March 24, 2017.

217 **"There's often two in a category":** Laurie Kretchmar, interview with the author, March 21, 2017.

218 **"those who thought the Web":** Janelle Brown, "What Happened to the Women's Web?"

218 **Canadian scholar and theorist Leslie Regan Shade:** Shade, "Gender and the Commodification of Community," 157.

218 **"There's the ultimate deception":** Francine Prose, "A Wasteland of One's Own," *New York Times,* February 13, 2000.

219 **Fucked Company posted a new notice:** http://web.archive.org/web/20001206145600/http://www.fuckedcompany.com:80.

219 **"with deep losses":** Larson, "Free Money."

219 **"peeling them off the ceiling":** Ibid.

220 **"It affected me":** McDaniel, interview with the author, March 9, 2017.

220 **Women.com's IPO was nowhere:** Myra M. Hart, "Women.com (B)," *HBS 9-802-109* (Boston: Harvard Business School Publishing, 2001), 1.

220 **Women.com even nosed past iVillage:** Larson, "Free Money."

220 **"Not my favorite topic":** McDaniel, interview with the author, March 9, 2017.

221 **"The Internet cannot sustain":** Jennifer Rewick, "iVillage.com to Buy Rival Women.com for $30 Million," *Wall Street Journal,* February 6, 2001.

CHAPTER THIRTEEN: THE GIRL GAMERS

222 **"Very early in life":** Jane Margolis and Allan Fisher, *Unlocking the Clubhouse: Women in Computing* (Cambridge, MA: MIT Press, 2002), 4.

222 **This discouragement permeates:** Sherry Turkle, *Life on the Screen: Identity in the Age of the Internet* (New York: Touchstone, 1995), 62.

223 **"It was the Andy Warhol period":** Brenda Laurel, interview with the author, August 9, 2016.

223 **"little-bitty feisty woman":** Ibid.
224 **"I had an epiphany":** Brenda Laurel, *Utopian Entrepreneur* (Cambridge, MA: MIT Press, 2001), 99.
224 **"pixels from Mars":** Laurel, interview with the author, August 9, 2016.
224 **"Without knowing it was hard":** "An Interview with Brenda Laurel (Purple Moon)," in *From Barbie to Mortal Kombat: Gender and Computer Games,* ed. Justine Cassell and Henry Jenkins (Cambridge, MA: MIT Press, 1998), 119.
226 **"You know what? I can't stand":** Laurel, interview with the author, August 9, 2016.
227 **"uncomfortable but fine, wild ride":** Brenda Laurel, *Computers as Theatre,* 2nd ed. (Boston: Addison-Wesley, 2014), 60.
227 **"When I interviewed men about VR":** Laurel, interview with the author, August 9, 2016.
228 **But just as Brenda was starting:** Akira Nakamoto, "Video Game Use and the Development of Socio-Cognitive Abilities in Children: Three Surveys of Elementary School Students," *Journal of Applied Social Psychology* 24 (1994): 21–22.
228 **"play games, to program":** Justine Cassell and Henry Jenkins, "Chess for Girls? Feminism and Computer Games," in *From Barbie to Mortal Kombat,* 13.
229 **"Mastery for its own sake":** "An Interview with Brenda Laurel (Purple Moon)," 122.
229 **"This soft mastery," she explained:** Turkle, *Life on the Screen,* 56.
229 **Interval spun Brenda's research:** Cassell and Jenkins, "Chess for Girls?," 11.
229 **"We cannot expect women to excel":** Sheri Graner Ray, *Gender Inclusive Game Design: Expanding the Market* (Hingham, MA: Charles River Media, 2004), 6.
230 **"the computer game equivalent of pink Legos":** "An Interview with Brenda Laurel (Purple Moon)," 122.
230 **"it's not only that the characters":** Ibid.
231 **"They truly helped teach me":** Kacie Gaylon, e-mail to the author, November 1, 2016.
231 **"Can you do this for boys?"** Laurel, *Computers as Theatre,* 172.
231 **"Some shit's going on":** Laurel, interview with the author, August 9, 2016.
232 **"find emotional resources within themselves":** Henry Jenkins, "'Complete Freedom of Movement': Video Games as Gendered Play Spaces," in *From Barbie to Mortal Kombat,* 285.
234 **"'virtuous cycle' where girls playing":** Misa, "Gender Codes," 13.
234 **"What Purple Moon and other 'girlie games'":** "Voices from the Combat Zone: Game Grrlz Talk Back," complied by Henry Jenkins, in *From Barbie to Mortal Kombat,* 330.
234 **"reinforce the very same stereotypes":** Rebecca Eisenberg, "Girl Games: Adventures in Lip Gloss," *Gamasutra,* February 12, 1998, www.gamasutra.com/view/feature/131660/girl_games_adventures_in_lip_gloss.php.
235 **"You can't get buy-in from somebody":** Laurel, interview with the author, August 9, 2016.
235 **"Paul Allen took us into Chapter 7":** Ibid.
235 **"But since Purple Moon did not make it":** Amy Harmon, "With the Best Research and Intentions, a Game Maker Fails," *New York Times,* March 22, 1999, www.nytimes.com/1999/03/22/business/technology-with-the-best-research-and-intentions-a-game-maker-fails.html.

235 **"Mattel was trying to protect their Barbie franchise":** Laurel, interview with the author, August 9, 2016.

236 **"We're always trying to heal something":** Laurel, *Utopian Entrepreneur,* 4–5.

EPILOGUE: THE CYBERFEMINISTS

237 **"THE CLITORIS IS A DIRECT LINE TO THE MATRIX":** Plant, *Zeros + Ones,* 59.

238 **"The technological landscape was very dry":** Virginia Barratt, e-mail to the author, December 1, 2014.

239 **"The Internet was far less regulated":** Francesca da Rimini, e-mail to the author, December 2, 2014.

239 **"As the population becomes widely familiar":** Scarlet Pollock and Jo Sutton, "Women Click: Feminism and the Internet," in *Cyberfeminism: Connectivity, Critique, Creativity,* eds. Susan Hawthorne and Renate Klein (North Melbourne, AUS: Spinifex Press, 1999), 38.

240 **"a question of survival and power":** Old Boys' Network, "Old Boys' Network FAQ," 2000, http://web.archive.org/web/20000424093036/http://www.obn.org:80/faq.htm.

240 **"Cyberfeminism only exists in the plural":** Cornelia Solfrank, "The Truth About Cyberfeminism," 1998, www.obn.org/reading_room/writings/html/truth.html.

240 **"not about boring toys for boring boys":** Old Boys' Network, "100 Anti-Theses," 1997. www.obn.org/cfundef/100antitheses.html.

240 **"virtual techno-paradise of the new millennium":** Renate Klein, "The Politics of CyberFeminism: If I'm a Cyborg Rather Than a Goddess Will Patriarchy Go Away?" in *Cyberfeminism,* 10.

241 **"Möbius strip of reality and unreality":** Katherine Cross, "Ethics for Cyborgs: On Real Harassment in an 'Unreal' Place," *Loading... The Journal of the Canadian Game Studies Association* 8 (2014): 4–21

241 **"We did what we had to do at the time":** Virginia Barratt, e-mail to the author, December 6, 2014.

Index

Page numbers in *italics* refer to photographs.